FORENSIC SPOOROLOGY: SEEING AND UNDERSTANDING HUMAN BEHAVIOR
THROUGH OBSERVATION, CLASSIFICATION, AND
INTERPRETATION OF SPOOR EVIDENCE

A thesis presented to the Faculty of the U.S. Army
Command and General Staff College in partial
fulfillment of the requirements for the
degree

MASTER OF MILITARY ART AND SCIENCE
General Studies

by

TYRON J. (TY) CUNNINGHAM
SUPERVISORY CRIMINAL INVESTIGATOR/CERTIFIED MASTER SCOUT TRACKER
UNITED STATES MARSHALS SERVICE
B.A., Social and Criminal Justice, Ashford University, Clinton, Iowa, 2009

Fort Leavenworth, Kansas
2011-01

REPORT DOCUMENTATION PAGE

Form Approved
OMB No. 0704-0188

Public reporting burden for this collection of information is estimated to average 1 hour per response, including the time for reviewing instructions, searching existing data sources, gathering and maintaining the data needed, and completing and reviewing this collection of information. Send comments regarding this burden estimate or any other aspect of this collection of information, including suggestions for reducing this burden to Department of Defense, Washington Headquarters Services, Directorate for Information Operations and Reports (0704-0188), 1215 Jefferson Davis Highway, Suite 1204, Arlington, VA 22202-4302. Respondents should be aware that notwithstanding any other provision of law, no person shall be subject to any penalty for failing to comply with a collection of information if it does not display a currently valid OMB control number. **PLEASE DO NOT RETURN YOUR FORM TO THE ABOVE ADDRESS.**

1. REPORT DATE (DD-MM-YYYY) 10-06-2011	2. REPORT TYPE Master's Thesis	3. DATES COVERED (From - To) AUG 2010 – JUN 2011

4. TITLE AND SUBTITLE	5a. CONTRACT NUMBER
Forensic Spoorology: Seeing and Understanding Human Behavior through Observation, Classification and Interpretation of Spoor Evidence	5b. GRANT NUMBER
	5c. PROGRAM ELEMENT NUMBER

6. AUTHOR(S)	5d. PROJECT NUMBER
Tyron J. (Ty) Cunningham, Supervisory Criminal Investigator/ Certified Master Scout Tracker	5e. TASK NUMBER
	5f. WORK UNIT NUMBER

7. PERFORMING ORGANIZATION NAME(S) AND ADDRESS(ES) U.S. Army Command and General Staff College ATTN: ATZL-SWD-GD Fort Leavenworth, KS 66027-2301	8. PERFORMING ORG REPORT NUMBER

9. SPONSORING / MONITORING AGENCY NAME(S) AND ADDRESS(ES)	10. SPONSOR/MONITOR'S ACRONYM(S)
	11. SPONSOR/MONITOR'S REPORT NUMBER(S)

12. DISTRIBUTION / AVAILABILITY STATEMENT
Approved for Public Release; Distribution is Unlimited

13. SUPPLEMENTARY NOTES

14. ABSTRACT

Humans have used traditional tracking "skills" to follow a quarry (human or animal) since the beginning of existence. "Visual Tracking, at its very basic level is the natural predatory hunting instinct of man." Even with this history, the sophistication of spoor (track, trackway) observation, classification, and interpretation has only recently within the last thirty years addressed the "why" of scientific enquiry as to human behavior and its influences on spoor evidence? Tracking "necessity" has decreased perceptually with the complexities of human movement patterns during the period of industrialization, but tracking "education" has increased reacquainting the professional fields of military, law enforcement, and search and rescue as to the who, what, when, and where, which is registered in any substrate (earth surface soil or vegetation) scene. This tracking education, although pervasive today, has not addressed in any sophistication the scientific processes of human behavioral influences on spoor evidence [cognitive behavioral brain-bound traits, gait sequential body-bound traits, weight convergence action-bound traits]. Thus, gait footfall sequences are encryptions that can only be exploited, analyzed, and disseminated properly by experienced trackers. This researcher has designed a modeling-based representation for overall decryption of primary movement patterns and secondary movement patterns that facilitates observation, classification, and interpretation of human behavior (locomotor and psychomotor) through spoor-chain signatures.

15. SUBJECT TERMS
Forensics, crime scene, human behavior, gait analysis, spoor analysis, man-tracking, track, sign, spoor

16. SECURITY CLASSIFICATION OF:			17. LIMITATION OF ABSTRACT	18. NUMBER OF PAGES	19a. NAME OF RESPONSIBLE PERSON
a. REPORT (U)	b. ABSTRACT (U)	c. THIS PAGE (U)	(U)	208	19b. PHONE NUMBER (include area code)

Standard Form 298 (Rev. 8-98)
Prescribed by ANSI Std. Z39.18

MASTER OF MILITARY ART AND SCIENCE

THESIS APPROVAL PAGE

Name of Candidate: Tyron J. (Ty) Cunningham

Thesis Title: Forensic Spoorology: Seeing and Understanding Human Behavior through Observation, Classification and Interpretation of Spoor Evidence

Approved by:

_____, Thesis Committee Chair
Mark M. Hull, Ph.D.

_____, Member
LTC Casey J. Lessard, M.A.

_____, Member
Kenneth J. Riggins, M.A.

Accepted this 10th day of June 2011 by:

_____, Director, Graduate Degree Programs
Robert F. Baumann, Ph.D.

The opinions and conclusions expressed herein are those of the student author and do not necessarily represent the views of the U.S. Army Command and General Staff College or any other governmental agency. (References to this study should include the foregoing statement.)

ABSTRACT

FORENSIC SPOOROLOGY: SEEING AND UNDERSTANDING HUMAN BEHAVIOR THROUGH OBSERVATION, CLASSIFICATION AND INTERPRETATION OF SPOOR EVIDENCE, by Tyron J. (Ty) Cunningham, 209 pages.

Humans have used traditional tracking "skills" to follow a quarry (human or animal) since the beginning of existence. "Visual Tracking, at its very basic level is the natural predatory hunting instinct of man." Even with this history, the sophistication of spoor (track, trackway) observation, classification, and interpretation has only recently within the last thirty years addressed the "why" of scientific enquiry as to human behavior and its influences on spoor evidence? Tracking "necessity" has decreased perceptually with the complexities of human movement patterns during the period of industrialization, but tracking "education" has increased reacquainting the professional fields of military, law enforcement, and search and rescue as to the who, what, when, and where, which is registered in any substrate (earth surface soil or vegetation) scene. This tracking education, although pervasive today, has not addressed in any sophistication the scientific processes of human behavioral influences on spoor evidence [cognitive behavioral brain-bound traits, gait sequential body-bound traits, weight convergence action-bound traits]. Thus, gait footfall sequences are encryptions that can only be exploited, analyzed, and disseminated properly by experienced trackers. This researcher has designed a modeling-based representation for overall decryption of primary movement patterns and secondary movement patterns that facilitates observation, classification, and interpretation of human behavior (locomotor and psychomotor) through spoor-chain signatures.

ACKNOWLEDGMENTS

My sincere thanks go to the members of my research committee for their extreme patience and guidance throughout my research. It is because of their insightful critical comments that I was able to stay on task while constantly framing and reframing my seemingly unquenchable linear thinking. Most important to my work was their untiring encouragement, which made my research joyful, exciting, and intellectually stimulating. I, therefore, have reached a higher intellectual level. Dr. Hull, in particular, made me realize the indivisible nature of the scientific method and the rule of law and the connectivity between the scientific method and empirical evidence. I am most grateful to have been supported by such a distinguished scholar and professor.

I would like to thank the U.S. Marshals Service for the selection to this renowned military college. Without the confidence of U.S. Marshal Mauri Sheer and Chief Deputy Tony Gasaway, I would not have been recommended to attend this remarkable institution. In addition, to those trackers I have been blessed to command on both the U.S. Marshals Tactical Tracking Unit (District of Alaska) and the U.S. Marshals Mounted Tracking Unit (District of Wyoming), remember always: "To the quarry lies the tracker and to the tracker lies the trail. There is always a trail." Lastly, yet most notable, to my family who paid the ultimate sacrifice of all--to live this year without a husband and father.

TABLE OF CONTENTS

ACRONYMS

ACBT	Action Bound Trait(s)
APEX	Apex of Foot Arc
AR	Alarm Reaction
ATP	Aerial Travel Point(s)
BGS	Behavior, Gait, Spoor
BLE	Baseline Establishment
BOBT	Body Bound Trait(s)
BRBT	Brain Bound Trait(s)
CSA	Collateral Spoor Area
DVO	Direct Visual Observation
EAT	Eight Adaptive Traits
ESA	Extended Spoor Area
FP	Flex Point
FPR	Fluctuating Pressure Releases
GFS	Gait-Footfall Sequencing
HAT	Head, Arms, Torso
HBI	Human Behavior Identification
IAT	Innate Adaptive Trait(s)
IMS	Imperturbable-Mind State
IP	Impact Point
IPG	Impact Point Gradient
IPR	Indicator Pressure Releases
IRA	Incident Reconstruction Area

LID	Landscape-Imposed Distortion
LKS	Last Known Spoor
LKT	Last Known Track
LMA	Loco-Motor Anomaly
LMBP	Loco-Motor Behavioral Programming
LMP	Loco-Motor Platform
MABT	Manifest Abdominal Trait
MACC	Macro Class Characteristic
MAIC	Macro Identifying Characteristic
MAT	Manifest Adaptive Trait(s)
MBS	Mind, Body, Spirit
MCIT	Manifest Cognitive/Intuitive Trait
MFYT	Manifest Force-Yield Trait
MID	Mechanical-Imposed Distortion
MIIC	Micro Identifying Characteristic
MIST	Manifest Imperturbability Mind/Steadfast Mind Trait
MOPT	Manifest Omni-Poise Trait
MRVT	Manifest Respiratory/Vocality Trait
MST	Manifest Synchronicity Trait
MVT	Manifest Volitional Trait
NVO	Non-Visual Observation
NWG	Normal Walking Gait
OCISE	Observation, Classification, Interpretation of Spoor Evidence
OSR	Observation and Spoor Recognition
PMA	Psycho-Motor Anomaly

PMBP	Psycho-Motor Behavioral Programming
PP	Pivot Point
PSA	Primary Spoor Area
SAL	Spoor Analysis Lane
SASA	Stride and Step Analysis
SCS	Spoor-Chain Signature
SID	Self-Imposed Distortion
SMS	Steadfast-Mind State
SSA	Secondary Spoor Area
TEC	Track Erosion Computation
TKR	Tracker
TP	Terminal Point
TPG	Terminal Point Gradient
TQR	Target Quarry Reference
TTP	Tactics, Techniques, Procedures
USF	Uniform Scale Format
VSP	Visual Search Pattern
WID	Weather-Imposed Distortion

ILLUSTRATIONS

TABLES

CHAPTER 1

INTRODUCTION

Forensics is the deliberate collection and methodical analysis of evidence that establishes facts that can be used to identify connections between persons, objects, or data.

> — Headquarters, Department of the Army
> ATTP 3-39.20, *Police Intelligence Operations*

Spoor is *a set of tracks laid on the ground* that are visible to a tracker. For example, "following the spoor." Spoor is totally interchangeable with the words tracks, trail, and set of prints.

> — David Scott-Donelan, *Tactical Tracking Operations:*
> *The Essential Guide for Military and Police Trackers*

Overview

This researcher has been tracking for four decades (see Appendix F for curriculum vitae). More than half of this time has been spent tracking as an enabler skill for professions within the federal government as a soldier, criminal investigator, and supervisory criminal investigator. Whether it was to follow the spoor-chain signature (footprint trail) of a quarry to close the time-distance interval for capture, or to follow the spoor-chain signature to establish the story of what happened at an urban or wilderness crime scene, it was only to assist the performance of primary duties as assigned. It was not until 1998, after nearly three decades of tracking as an enabler skill, that this researcher was given the assignment to develop and command two tracking units for the U.S. Marshals Service as Chief of Scouts: The District of Alaska Tactical Tracking Unit (TTU) from 1998 to 2002 and the District of Wyoming Mounted Tracking Unit (MTU) from 2003 to 2006.

It was important in the development of these units to define the mission, the scope of work, and the product for tracker case development; therefore, assisting local, state, and federal law enforcement agencies in criminal prosecutions. A thorough canvas of available regulatory tracking standards was researched before the TTU became operational. This research gleaned that there was no regulatory tracking standard because there was no self-regulating body of professional trackers whom by education, training, and experience had come together to establish an organization (such as the International Association of Identification or American Society of Questioned Document Examiners) that the rest of the world could look to for professional tracking standards. This researcher then turned to individual private tracking companies who contracted tracking training and services to individuals and governments. This data established some common norms for tracking but the prevalent norm overall was that most of these companies and individuals had different standards for tracker skill, performance, and certification.

As the development of TTU progressed, certification became the focus so that tracker standardization could be relied on by the courts for viability and reliability. The tracker certification used by the TTU, and subsequently by the MTU, was based on the defined mission, scope, and culmination of all available tracker standard data available at the time. Yet, this was for internal use by the U.S. Marshals Alaska TTU and Wyoming MTU for those particular districts, courts, and did not establish a universal tracker standard from which all courts could rely on as a universal tracker skill standard. To this date, this researcher has gone through the approved certification process ratified by the

U.S. Marshal and Chief Deputy U.S. Marshal for both the District of Alaska and District of Wyoming achieving in 2004 the award of Certified Master Scout Tracker (CMST).

With this background tracking knowledge and expertise, this researcher begins the exploration into an area of tracking research that is held to be suspicious by other professionals outside of tracking, yet the suspicion is purely due to lack of education and experience. It is the hope of this research thesis to lay the foundation for the science of tracking (spoorology), its applicability as "scientific, technical, and specialized" knowledge for the courts, open academic and tracker dialogue at all levels, and overall further scholarly tracking research so that tracking can find its place among the other established sciences of our day.

Spoor Terminology

Every science is defined by the language it uses and this is true of tracking as well. In researching all available sources of tracking knowledge to glean the language most common to all (see Appendix E), it became apparent that this tracking language was similar in most respects between all schools of thought except for the term "spoor," which is used primarily by those trained through tactical and combat tracking schools. The step-by-step schools do not use this term, instead using the term track, trail, and footprint. Both schools of thought are accurate.

A general definition of the term spoor refers to "a track, a trail, a scent, or droppings especially of a wild animal" or "a trace by which the progress of someone or something may be followed."[1] In noun form, spoor would refer to "the trail of an animal or person, esp as discernable to the human eye."[2] In verb form, spoor would be "to track

3

. . . by following its trail."[3] The term spoor was first known to be used in 1823 and stems from Dutch and old English.[4]

Spoor, as far as this researcher can ascertain, is the only word in the current English language that is tracking specific. The word spoor stands alone in use. Terms like tracking, track, sign, trail, and footprint are often times used in other contexts like computer tracking, mail tracking, track button, track and field, street sign, trail of tears, tent footprint, footprint of a building, and others. These contexts have confused the use of the term tracking, in its original context, and therefore, require predicated terms like traditional and primitive to refer to the following of an animal or human tracks. For this researcher, this thesis uses the spoor terminology to isolate the science of tracking (spoorology) from the ambiguity of word usage in other contexts and includes many new terms specific to this thesis on forensic spoorology.

Thesis Exploration

This thesis explores the topic of human behavioral influences on spoor evidence and the intrinsic observation, classification, and interpretation of spoor evidence (OCISE) modeling based on human movement patterns that are left as signature stories imprinted in the substrate (earth's surface soil and vegetation). These project walking gait encryptions that can only be analyzed properly by experienced trackers. This researcher has designed a modeling-based representation for overall decryption of human movement that facilitates recognition, classification, and interpretation of people and their behaviors by their footfall sequences. The OCISE method encompassing a "totality of the circumstances" (also known as the "sign story"[5]) approach is designed to answer the interrelated questions: Is human behavior recognizable in the gait footfall sequences and spoor-chain signature? and How much locomotor and psychomotor information is contained in the collection of spoor evidence? The spoor evidence data is tested in this study by direct visual observation within spoor analysis lanes and confirmed in incident reconstruction areas to validate collected research data based on the Behavior, Gait, Spoor (BGS) Paradigm (figure 1).

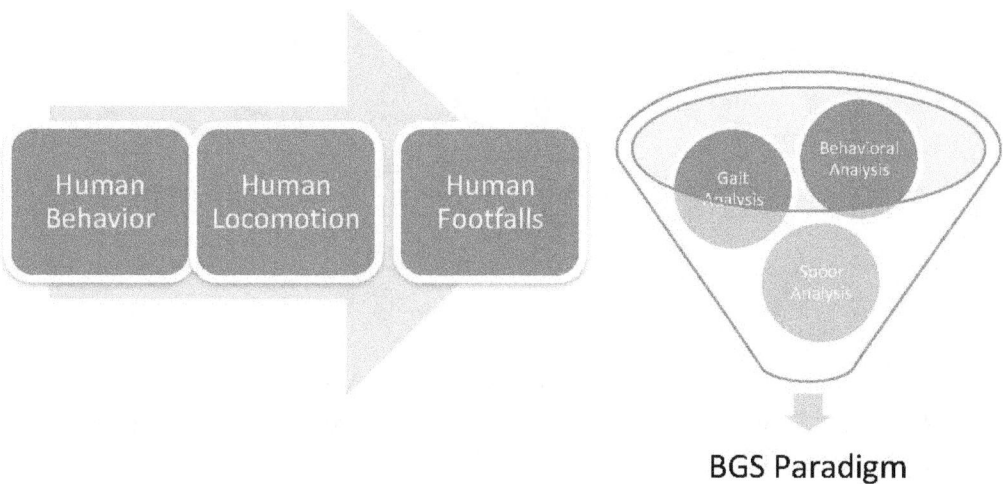

Figure 1.   Behavior, Gait, Spoor Paradigm
Source: Created by author.

<u>Establishing the Behavior, Gait, Spoor Paradigm</u>

The scientific community as well as science educators have used and taught about the laws of motion (see figure 2) established by the well-known British physicist Sir Isaac Newton since they were established as scientific adages not to be questioned.[6]

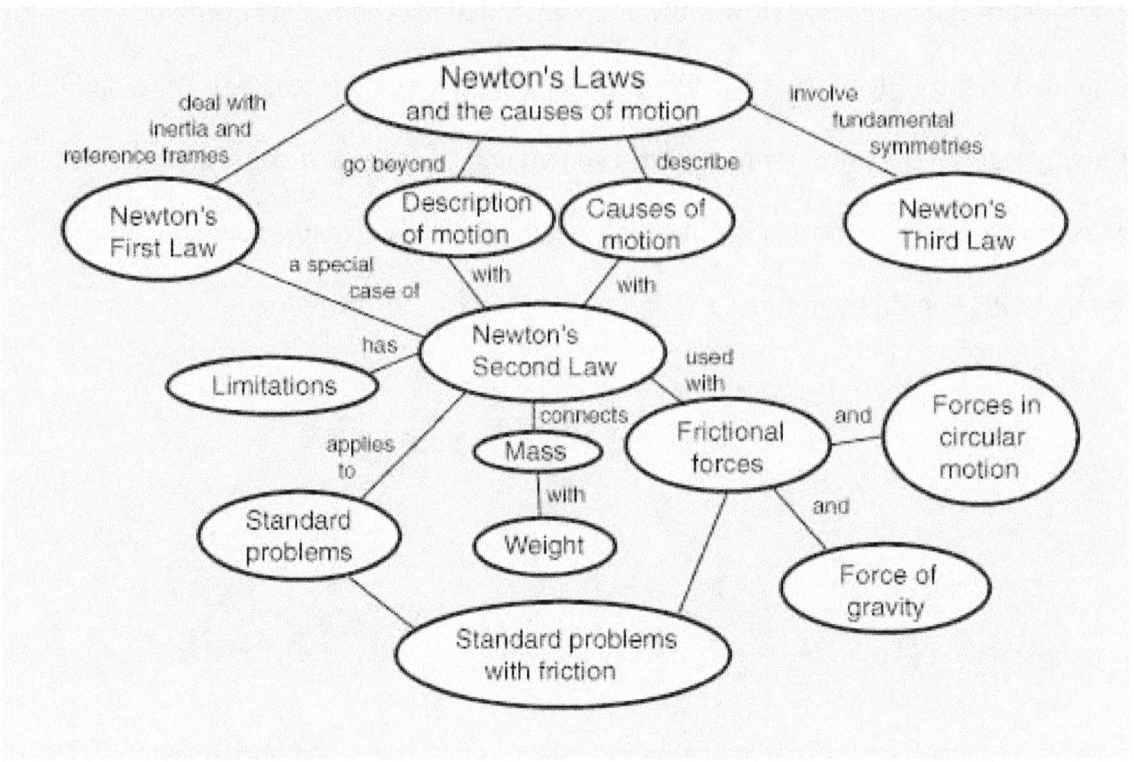

Figure 2.   Newton's Laws of Motion
Source: C. R. Nave, "Laws of Motion," 2010, http://hyperphysics.phy-astr.gsu.edu/
hbase/newt.html#nt2cn (accessed 13 April 2011).

They are used as the basis for the BGS paradigm:

1. Every body (locomotor) continues in its state of rest of uniform rectilinear

   motion unless compelled to change its state by the action of forces

   (psychomotor).

2. The change of motion (register deviation anomaly) is proportional to the force

   acting, and takes place along the straight line along, which the force acts.

3. There is always a reaction equal and opposite to action (psychomotor); or, the

   actions of two bodies (locomotor and psychomotor) on each other are always

   equal and opposite.

Newton's linear paradigm has continued as an important implement from, which to understand, visualize, describe, and control nature. It contains implications about the elementary symmetry of the universe in that a state of motion in a straight line must be just as natural as being at rest. Both of, which, require a force to action. It also contains the physical occurrences, which changes motion in proportion to the force applied against it. Lastly, that all forces in the universe happen in equal but oppositely directed pairs upon the body. Therefore, there are no isolated forces, in that, for every external force that acts on the bodies system there is a force of equal magnitude but opposite direction, which acts back on the bodies system, which exerted the external force. In looking at internal forces, a force on one part of bodies system will encounter by a reaction force on another part of the bodies system so that an isolated body system cannot by any means exert a net force on the bodies system as a whole. In other words, the body as a system cannot move itself into motion with purely internal forces. To achieve a net force and an acceleration, the bodies system must interact with objects external to itself to achieve linear motion.

The basic attributes of the linear paradigm are derivative of the concepts contained within the laws of motion:[7]

1. The output actions of a linear system are directionally proportional to the input to the system. For instance, the human bodies' actions (output) to move is directly equal to the force to move (input) applied through reaction. Alternatively, the power of the whole body is directly proportional to the capabilities of all the parts of the body. Therefore, external pressure from the environment changes the mind's intent, which is gradually adjusted to

compensate for the level of environmental pressure imposed on the mental will.

2. All linear actions submit to the rule of additivity. If a system is disturbed by two inputs from A and B, the outputs of both A and B when summed together are equal to the outputs of both if they were reversed as in B and A. If through human locomotion the body moves 100 steps, stops for a time, and then moves 100 more steps later, by applying linearity, the total results of those two subsequent yet separate human movement patterns is the same should the human movement initially move 200 steps with no stop. Therefore, when the human body moves through locomotion, the gait steps no matter the pattern, are sequential as in forward propulsion thus establishing the gait-footfall sequence. This human movement pattern through gait must interact with the environment by contacting the substrate through the laws of motion and through the rule of linearity, which establishes the spoor-chain signature.

3. The ratio of output of actions to input of forces must continue to be constant. Therefore, all future actions and behavior can be forecasted through the method of extrapolation. If through the output of human movement contained in the rule of additivity and established through both gait-footfall sequences and spoor-chain signatures through the input of environmental forces contained in proportionality. Then, the baseline capsulated the extrapolation forecast within all human locomotion registered in any substrate.

4. A linear system is very conclusive in nature. Given a set algorithm, which establishes a set of applicable rules makes future outcome deterministic.

Therefore, just as the baseline complies with linearity. So to, does the human movement patterns for each of the human behaviors comply with linearity once they are recognized in the gait-footfall sequence and spoor-chain signature. This base and underlying assumption of linearity is fundamental to the forensic sciences.

The BGS paradigm, in its construction, identifies and coexists with the linear paradigm of, which Newton asserts. The human body will not move until a force is applied against the bodies resting state and once the body is in motion it will continue in that linear motion until affected by force. The very force applied is the internal cognition and volition of the brain to force the motion of the body to action. This expresses the minds will through motion, which produces relevant body motions including gait and the subsequent gait-footfall sequence of the spoor-chain signature.

The linear paradigm is authoritative and signifies the importance of the BGS paradigm to the OCISE algorithm. Linearity gives BGS its footing. The spoor that is registered in a substrate is the power of nature's forces upon the human body as external forces require the body to actions, which maintain proportionality. The human body is constantly in rectilinear motion in one form or another. This movement is unique in its pattern to the trackmaker and establishes the gait-footfall sequence and its subsidiaries of movement implicating the mind, which produced the intent of the movement to gain proportionality, additivity, and output to input constancy, with conclusive and deterministic extrapolations. With the BGS paradigm, and through the OCISE algorithm, human behavior identification is reliable when recognized in the spoorological context.

Spoorology as Science

As quoted at the beginning of this chapter, "Forensics is the deliberate collection and methodical analysis of evidence that establishes facts that can be used to identify connections between persons, objects, or data."[8] Continuing this line of thought, Richard Saferstein states that; "Forensic science is the application of science to the criminal and civil laws that are enforced by police agencies in a criminal justice system."[9] Forensic science work is best understood when the topic of the crime laboratory is invoked. Yet, forensic science in the law enforcement sense is a force multiplier because it is an important aid to a criminal investigation, and therefore, requires the involvement of scientific, technical, and specialized knowledge of, which a spoorology (behavior, gait, and spoor) analysis falls. In fact, Louis Liebenberg in his book *The Art of Tracking: The Origin of Science* asserts that the methods of tracking, which date back to the first hominids establish the origins of scientific thought, which we use today in the scientific community.

> And while the similarities between tracking and modern science may suggest how science originated by means of biological evolution, the differences between them may give some indication of how science subsequently developed by means of cultural evolution…As perhaps the oldest science, the art of tracking is not only of academic interest, it may also be developed into a new science with many practical applications.[10]

Further Liebenberg shows in nature conservation how tracking is used to effectively control the crime of poaching. Through implementation of tracker skills, often poachers are able to be intercepted before they can do any damage to wildlife. Trackers can find signs of poaching and follow the spoor to apprehend the guilty individuals. Also following poaching spoor, trackers are even able to identify individual poachers by their spoor.[11]

To be "spoor conscious" opens up a world, which cannot be seen by the untrained eye. Most criminals move and act when no one is watching. Spoorology is the science to see and interpret the spoor evidence. When a tracker looks and sees spoor in the substrate of earth's surface, the story of the unseen workings of animal and man are seen and understood. By the reconstruction of movements from footprints, the tracker visualizes the trackmaker and actual sees them leaving the spoor-chain signature. This is the way, to the tracker; the whole story is unfolded as to what has happened when no one was there to see a criminal at work. The crime scene comes alive with empirical spoor data!

It is common knowledge within the community of trackers, that tracks give a personal account of the undisturbed everyday lives of all trackmakers, both human and animal. The tracker interpretation of spoor is based off a unique understanding of the behavior of the trackmaker. This understanding of behavior for the tracker (spoorologist) is different from the understanding of behavior by a psychologist, anthropologist, or zoologist. A tracker sees the behavior of a human or animal through a different prism of knowledge, which must be understood by the other sciences in order to understand the trackers interpretations, which, as described earlier, come through proportionality, additivity, output and input constancy, and conclusive and deterministic extrapolations. This thesis will give this scientific understanding.

Tracking as a forensic modality is concerned with scientific accuracy. Both the BGS paradigm and the OCISE algorithm, which is built off the paradigm shows the human patterns, which are left in the substrate, thus fulfilling by definition what forensics analysis is. Forensics as explained further in *Police Intelligence Operations:*

11

It is most commonly associated with evidence collected at crime scenes or incident sites. . . . Forensics is typically employed to support legal proceedings that lead to criminal prosecution. . . . Forensic analysis expands the ability of police…to establish trends, patterns, and associations by providing scientific documentation of relationships between persons, objects, or data. Criminals, terrorists, or other threat elements tend to operate in predictable ways. The challenge is identifying the patterns.[12]

Tracking is first a specialty skill in observation of physical evidence, which is not seen by most criminal investigators, crime scene technicians, and forensic scientists. Footwear impression evidence examiners cannot even see all the spoor evidence that is left within the crime scene. They only examine the obvious spoor, which is registered in soft ground or in snow by isolated tracks. This visibility gap in knowledge leaves an examiners eyes outside the realm of linearity. One of the leading footwear impression evidence examiners, William Bodziak, former FBI agent stated in his book *Footwear Impression Evidence: Detection, Recovery, and Examination*, "What is not looked for will not be found."[13] Dwayne Hildebrand in his article "Footwear, The Missed Evidence," explains the two reasons footprints are overlooked:[14]

1. The lack of training and education in the proper searching, collection and preservation of the evidence and;

2. The evidence is undervalued or not understood.

Bodziak's statement above supports why other sciences cannot see all the spoor evidence at crime scenes.[15] This counter statement for the science of spoorology is salient and accurate: "What is looked for will be found." The concept that only if you see it, does it exist" is the scientific method. Therefore, this thesis will demonstrate the science of spoorology, its admissibility as testimony in court, and the algorithm, which frames spoorology use of the BGS paradigm with roots in linearity and laws of motion.

Spoorology and Admissibility In Court

In 1923, in the District of Columbia Circuit Court, the precedence was set for admissibility of scientific evidence in the court of law. In *Frye v. United States*, the court ruled as follows:

> Just when a scientific principle or discovery crosses the line between the experimental and demonstrable stages is difficult to define. Somewhere in this twilight zone the evidential force of the principle must be recognized, and while the courts will go a long way in admitting expert testimony deduced from well-recognized scientific principle or discovery, the thing from, which the deduction is made must be sufficiently established to have gained general acceptance in the particular field in, which it belongs.[16]

For years the *Frye* standard was the practice in court where the idea of what the court called the "generally accepted" procedure, technique, or principle had to meet this standard. The general acceptance standard could be established through written books and papers written on the subject, as well as prior judicial decisions relating to the reliability and acceptance of the procedures.[17] The problem inherent in this standard is the continuing practice in the scientific community to research and produce new and novel science, which is not yet generally accepted within the scientific community. For this purpose, many courts, as an alternative to the *Frye* standard, believe that the Federal Rules of Evidence contains language, which advocates a more flexible standard that does not rely on general acceptance as established in *Frye*.[18]

In Rule 702 of the Federal Rules of Evidence, which covers the testimony of experts, it reads:

> If scientific, technical, or other specialized knowledge will assist the trier of fact to understand the evidence or to determine a fact in issue, a witness qualified as an expert by knowledge, skill, experience, training, or education, may testify thereto in the form of an opinion or otherwise, if (1) the testimony is based upon sufficient facts or data, (2) the testimony is the product of reliable principles and

methods, and (3) the witness has applied the principles and methods reliably to the facts of the case.[19]

This delineates a broad application of expert testimony for admissibility so long as the trial judge determines the necessity of the expert testimony to the triers of fact. The fields of knowledge, which this rule applies is not limited merely to the "scientific" and "technical" but extends to all "specialized" knowledge. Similarly, the expert is viewed, not in a narrow sense, but as a person qualified by "knowledge, skill, experience, training or education." Therefore, within the latitude of the rule exists not only the experts in the strictest sense of the word, but also the larger group called "skilled" witnesses.[20]

To substantiate the Federal Rules of Evidence, the U.S. Supreme Court in *Daubert v. Merrell Dow Pharmaceuticals, Inc.*[21] asserted in their ruling in 1993 that *Frye* is not the absolute prerequisite to the admissibility of scientific evidence under the Federal Rules of Evidence. The court held that:[22]

> (a) Frye's "general acceptance" test was superseded by the Rules' adoption. The Rules occupy the field, United States v. Abel, 469 U.S. 45, 49 , and, although the common law of evidence may serve as an aid to their application, id., at 51-52, respondent's assertion that they somehow assimilated Frye is unconvincing. Nothing in the Rules as a [509 U.S. 579, 2] whole or in the text and drafting history of Rule 702, which specifically governs expert testimony, gives any indication that "general acceptance" is a necessary precondition to the admissibility of scientific evidence. Moreover, such a rigid standard would be at odds with the Rules' liberal thrust and their general approach of relaxing the traditional barriers to "opinion" testimony.

> (b) The Rules-especially Rule 702-place appropriate limits on the admissibility of purportedly scientific evidence by assigning to the trial judge the task of ensuring that an expert's testimony both rests on a reliable foundation and is relevant to the task at hand. The reliability standard is established by Rule 702's requirement that an expert's testimony pertain to "scientific . . . knowledge," since the adjective "scientific" implies a grounding in science's methods and procedures, while the word "knowledge" connotes a body of known facts or of ideas inferred from such facts or accepted as true on good grounds. The Rule's requirement that the testimony "assist the trier of fact to understand the evidence or to determine a fact

14

in issue" goes primarily to relevance by demanding a valid scientific connection to the pertinent inquiry as a precondition to admissibility.

(c) Faced with a proffer of expert scientific testimony under Rule 702, the trial judge, pursuant to Rule 104(a), must make a preliminary assessment of whether the testimony's underlying reasoning or methodology is scientifically valid and properly can be applied to the facts at issue. Many considerations will bear on the inquiry, including whether the theory or technique in question can be (and has been) tested, whether it has been subjected to peer review and publication, its known or potential error rate and the existence and maintenance of standards controlling its operation, and whether it has attracted widespread acceptance within a relevant scientific community. The inquiry is a flexible one, and its focus must be solely on principles and methodology, not on the conclusions that they generate. Throughout, the judge should also be mindful of other applicable Rules.

The flexibility of *Daubert* in the case of *Kumho Tire Co., Ltd. v. Carmichael*,[23] is solidified by the court when they ruled in 1999 that:

This language makes no relevant distinction between "scientific" knowledge and "technical" or "other specialized" knowledge. It makes clear that any such knowledge might become the subject of expert testimony. In *Daubert*, the Court specified that it is the Rule's word "knowledge," not the words (like "scientific") that modify that word, that "establishes a standard of evidentiary reliability." 509 U.S., at 589--590. Hence, as a matter of language, the Rule applies its reliability standard to all "scientific," "technical," or "other specialized" matters within its scope. . . . Neither is the evidentiary rationale that underlay the Court's basic *Daubert* "gatekeeping" determination limited to "scientific" knowledge. *Daubert* pointed out that Federal Rules 702 and 703 grant expert witnesses testimonial latitude unavailable to other witnesses on the "assumption that the expert's opinion will have a reliable basis in the knowledge and experience of his discipline." *Id.*, at 592 (pointing out that experts may testify to opinions, including those that are not based on firsthand knowledge or observation). The Rules grant that latitude to all experts, not just to "scientific" ones.[24]

A case that exemplifies the type of flexibility in the use of scientific, technical, and other specialized knowledge is *Henry v. Ryan, et al.* a 1986 murder case in the State of Arizona. Here Petitioner Graham Henry in 2009, a state prisoner under sentence of death, filed an Amended Petition for Writ of Habeas Corpus alleging that he is imprisoned and sentenced in violation of the United States Constitution. The petition raises twenty-five claims of, which evidence expansion was one. In reviewing the underlying allegations on, which several of Henry's claims are founded, the court established that the,[25]

> Petitioner testified that he was asleep in the camper of the victim's truck while Foote drove and Estes occupied the passenger seat. He awoke to hear Foote and Estes arguing as the truck turned onto a rough side road. When the truck stopped, Petitioner exited from the rear of the vehicle and saw Foote dragging Estes backwards up a berm alongside the road. By the time Petitioner exited the rear of the camper, Foote had stabbed Petitioner then leaped across the berm and ran to Estes, attempting to revive him and then dragging him to the location where the body was found.

> Bernell Lawrence, the State's expert tracker, examined the crime scene photographs taken by Detective Patterson on June 8, 1986; he also visited the scene on June 14. Lawrence testified that the evidence showed that Estes was led up the roadside berm by one or more people; that he was then dragged face-down away from the berm by a person wearing boots (Petitioner) and a person wearing flat-soled shoes (Foote), one on either side; that he was stabbed to death at the location to, which two people had dragged him; that the person wearing boots dragged the body to another location underneath a bush and then walked back to the vehicle; and that the person wearing the flat-soled shoes took the victim's walker a short distance away and threw it into the desert.

The petitioner in this case wanted to discredit the testimony of the State's tracking expert by introducing evidence that was excluded by the trail judge. The federal judge reaffirmed the rulings of the trail court and for purposes of the petition and claims relied on the tracking evidence produced at trail by the state's expert tracker. The knowledge of the tracker in this case was needed by the trial court as specialized and the flexibility of the court in this tracking evidence's admissibility met the standard to allow the evidence to the triers of the facts.

As stated earlier, Rule 702 of the Federal Rules of Evidence as well as case precedence asserts the use of scientific, technical, and specialty knowledge of, which spoorology (tracking evidence) assuredly is. This thesis will establish spoorology as scientific knowledge based on an assessment of whether the testimony's underlying reasoning or methodology is scientifically valid and properly can be applied to the facts at issue based on Rule 104(a), which includes:

1. Whether tracking technique can be tested.

2. Whether tracking has been subjected to peer review and publication.

3. Tracking known or potential error rate and the existence and maintenance of standards controlling its operation.

4. Whether tracking has attracted widespread acceptance within a relevant scientific community.[26]

This standard set by the Federal Rules of Evidence through judicial inquiry is a flexible one, and the focus of the thesis on Forensic Spoorology must be based solely on reliable principles and sound methodology.

Crime Scene Evidence

There are a variety of ways that law enforcement agencies identify persons based on their biometric characteristics, such as with fingerprinting, iris scans, voice analysis, and handwriting analysis.[27] Although these methods are sufficient to isolate identity, many criminals realize this by changing their methods of operations.[28] For instance, wearing gloves so as not to leave fingerprints, or changing their handwriting to disguise a message, or voice alterations both intended and unintended. Iris scans are purely to authenticate identity and cannot as of yet be used to infer probable cause that a crime was committed and who the suspect might be.[29] This indeed causes inherent problems in the collection and analysis of crime scene evidence. From this problem-set came the science of footwear impression evidence, which collects an unknown footprint in order to match it with a known suspect's shoe.[30] Now, criminals are beating this evidence by destroying

17

the shoes worn while committing crimes.[31] Such problems can be mitigated by tracking

data (their human movement pattern sequences left in the substrate) collected at the crime

scene and interpreted to glean spoor evidence. It is therefore necessary to develop a

method, which if applied, can register a suspect's psychomotor primary movement

patterns and secondary movement patterns as well as an algorithm for analyzing these

movements to detect whether two given datasets belong to the same person and what

human behaviors are recognized by the datasets.

Gait and Behavioral Analysis

Before delineating the range and scope of spoor analysis (tracking) historically

and practically, a brief understanding of gait and behavior is essential to begin any spoor

study. Normal walking gait is the forward propulsion of a person that is determined by an

individual's height, weight, limb length, foot length, footwear, and personal

characteristics of motion.[32] Therefore, from a spoor locomotor behavioral perspective,

gait is used to recognize known (OCISE) persons and to classify unknown (OCISE)

persons (see figure 2). From a spoor psychomotor behavioral perspective, gait extends to

include the head, arms, and torso.[33] Posture of the head, arms, and torso in relation to the

locomotor apparatus and the mental processes that influence and determine (output of

behavior) where the feet are placed, as the mind processes how to move in and through

an environment, is also critical to human locomotion and spoor evidence and is

synchronized through the eight adaptive traits based on linearity and laws of motion.[34]

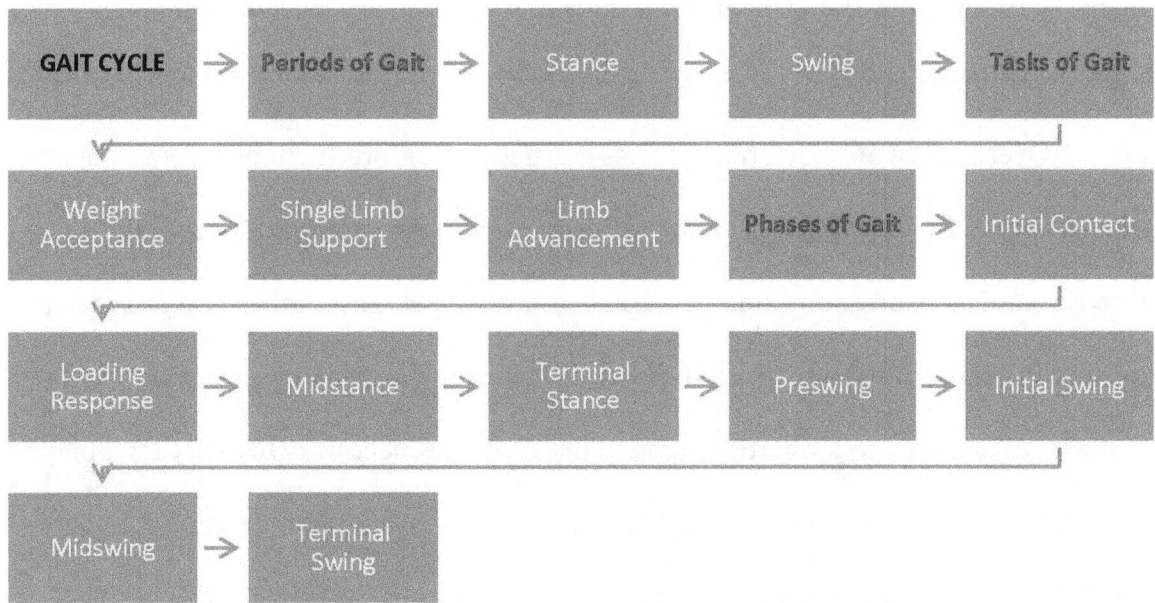

Figure 3.　Gait Cycle of Human Locomotion
Source: Created by author using data from Ed Ayyappa "Normal Human Locomotion, Part 1: Basic Concepts and Terminology, *American Academy of Orthotists and Prosthetists* 9, no. 1 (1997): 10-17, http://www.oandp.org/jpo/library/print Article.asp?printArticleId=1997_01_010 (accessed 13 April 2011).

Tracking and its Relationship to Spoorology

People of many cultures have used tracking for a long time.[35] The art of tracking was originally used to find sustenance.[36] Animal spoor (tracks, signs or trails) told people what animals were in the area and then hunters knew where to hunt. People would also set traps where spoor indicated animals passed regularly. Hunters also used tracks to locate their wounded target.[37] These applications of traditional tracking are still used today, but tracking has expanded into other mission specific areas based on other salient needs such as law enforcement, military, and search and rescue.[38] The first major contributor of trackers in police work within the United States came from the United

States Border Patrol whose responsibility it is to locate and apprehend illegal aliens who enter the United States.[39]

Over the last 30 years, tracking has assisted law enforcement officers throughout the United States in their duties to include patrol, investigations, crime scene, and forensics.[40] The number of trained trackers has grown, yet most of these trackers focus their skills on only observation and cursory classification in more of a "primitive" context. This has led to a significant gap in the skill level necessary to apply tracking in the scientific methodologies of observation, classification, and interpretation for court reliability. In an effort to close that gap, this research has led to delineated spoorological data definition, measurement, and analysis in a scientific systems approach to the component parts working within the framework of "totality of the circumstances" and applying this study of spoor to the law.

Spoorology's Functional and Purposeful Design

Forensic spoorology is application of advanced tracking skills to collect track evidence to apply it in a court setting. Tracking is following someone by stringing together a continuous chain of his or her spoor.[41] Spoorology (human behavior, locomotion, and spoor), then is the scientific study of spoor. Spoor evidence comes from the human body's functional task of maintaining kinetic stability during locomotion, which during forward progression transfers weight, which either foot must accept, before limb support or limb advancement can occur in succession. These footfall impressions are what every tracker observes, classifies-and with advanced scientific study-interprets to glean the behaviors of a quarry.[42]

Tracking, for forensic purposes, has been successful on a limited basis by police trackers for many years, but police administrations, attorneys, forensic and crime scene professionals have not grasped how useful tracking is to the fact-gathering process because of tracking's misinterpreted methodologies, which these professionals see as non-scientific.[43] The courts, however, have recognized tracking as a useful specialized knowledge to help the finders of fact as well as the court in understanding spoor evidence.[44] Yet, to this day, tracking is not widely accepted as a science. Those not familiar with tracking (this unfortunately is most people) usually think of this area of discussion as a supernatural ability to see the markings left where human or animal have walked, let alone understand the mental state and behaviors associated with locomotor and psychomotor human movement. This perception is truly limited in scope, yet casts shadows of doubt on tracking's applicability in our day as a scientific methodology, which is reliable, creditable, and can provide the necessary scientific accuracy to the fact finder.

Today, with the proper training, police trackers and investigators can observe, classify, and interpret spoor evidence to locate fugitives, find lost persons, analyze crime scenes, and testify in court to the spoor evidence recovered. Fish and Wildlife trackers can use tracking to find poachers and use spoor to learn how and why animals were killed. They also can use tracking to determine where a person trespassed onto private property and what they did while they were there.[45] In other situations, patrol officers investigating an abandoned vehicle can use tracking to see the movements the occupant made around the vehicle and determine where the person went.[46] In search and rescue operations, tracking can be used to locate missing persons. Tracking is also used to locate and keep under surveillance clandestine drug labs or marijuana fields.

Due to the limited amount of written material, it is difficult to inform people about forensic use of tracking, tracking's relevance to police work, and its accuracy. Neither articles, books, nor the research literature available provide information with, which to fully explore the accepted use of tracking within the police profession as related to the law. However, there is significant information available on tracking humans in general. The public and the police have primarily learned about tracking from its portrayal on radio and television. The media reports on tracking focus on its use in major publicized manhunts such as the four corners manhunt in Utah.[47] As a result, people do not realize it can be used for many other police purposes besides hunting fugitives. In addition, fictional writing about tracking creates more false assumptions. The most accurate information comes from trackers who use the skill as a practice because they understand the subject matter thoroughly and can provide information from their experiences from, which specific and significant detail can be gleaned. This thesis will provide analysis to the tracking modalities of this forensic science giving definition to methods of observation, clarity to methods of classification, and understanding to methods of interpretation as to spoor evidence.

Significance of the Research

Tracking is a beneficial skill of practical use and has been since the beginning of hominid existence.[48] It is still applicable today as a fundamental skill within the hunter-gatherer societies around the globe. Tracking is an enabler skill that is used to enhance the needs of many other sciences like geology, anthropology, archeology, paleontology,

22

etc. As described earlier, Liebenburg suggests that the art of tracking is the origin of science.[49] Relevant to the longevity of the tracking skill and with the need to clarify tracking's practicality as the forensic science of spoorology today, this study will answer three key questions: (1) whether the spoor evidence collected through observation, classification, and interpretation of spoorological research will establish the scientific accuracy requirements for court viability and reliability; (2) whether or not with this collected research can this methodology include or exclude suspects, and in turn, help the criminal justice system work more effectively; and (3) is human behavior recognizable in the spoor-chain signature?

Limitations

Due to the inherent complexities of this particular subject, it is beyond the scope of this research to discuss every tracking tactic, technique, and procedure (TTP). The research will be focused on the forensic modalities of tracking contained in the science of spoorology, namely the OCISE (see figure 3). OCISE allows civilian and military police evidence gatherers the best applicable practices of observing spoor in any environment and circumstance, classifying spoor into essential formulas for recovery and capture-ability of locomotor behavior, and interpreting spoor to examine psychomotor behavior necessary to understand quarry mental state and relevant ideo-motor actions. Discussion is also then limited to collateral research into human locomotion and gait science; tracking fundamentals and pursuit applications; trace evidence and footwear impression evidence; and case law as to admissibility of spoor evidence in court. The number one overriding limitation through the research venue is that no scientist has conducted this particular study in reference to spoor evidence viability and reliability for court admissibility before, so there remains a need to study human behavioral trait recognition comprehensibly of which this study attempts to do through the OCISE algorithm.

Assumptions

This researcher assumes that the spoor evidence collected from this research will glean both locomotor and psychomotor human behavior, and are therefore, accepted as fact so that a thorough focused analysis can be conducted into the salient parts. This researcher also assumes that the practice of the science of spoorology will better preserve evidence that is currently damaged and destroyed within any given crime scene because that which is not looked for will not be found.[50] Although, Forensic Spoorology is a relatively new science[51] and this is the first study into its practicalities and usefulness, the general subject of tracking, however, has little new information available. This thesis addresses a practical approach to OCISE, which can be applied in any circumstance to see and understand spoor and related behavior more efficiently. However, it does take applied research to gain extensive experience in application of OCISE algorithm methodologies in order to gain high levels of skill accuracy.

Organization

This thesis contains five chapters. Chapter 1 is an introductory overview of the science behind tracking, the legal admissibility of tracking as evidence, and the historical background of spoor and tracking evidence. Chapter 2 discusses the available information concerning tracking contained in various literature and other sources: books, magazines, journals, government documents and publications, DVD and VHS sources, newsletters, and electronic media. Chapter 3 discusses the applicable methodology used by this researcher. Chapter 4 analyses the primary and secondary research questions based on this researchers spoor study. It presents OCISE as a viable and reliable means for observing, classifying, and interpreting spoor evidence and recommends this as a practical science to evaluate the spoor evidence at a crime scene and therefore to assign locomotor and psychomotor behavioral actions to the spoor. Chapter 5 justifies this researcher's analysis with conclusions and recommendations.

---

[1]Merriam-Webster.com, "Spoor," http://www.merriam-webster.com/dictionary/spoor (accessed 28 April 2011).

[2]The Free Dictionary.com, "Spoor," http://www.thefreedictionary.com/spoor (accessed 28 April 2011).

[3]Ibid.

[4]Merriam-Webster.com.

[5]Joel Hardin and Matt Condon, *Tracker: Case Files & Adventures of a Professional Mantracker* (Everson, WA: Joel Hardin Professional Tracking Services), 256.

[6]Dennis T. Gyllensporre, "Adding Nonlinear Tools to the Strategist's Toolbox" (Master's Thesis, U.S. Army Command and General Staff College, 2001), 2.

[7]Linda P. Beckerman, *The Non-Linear Dymanics of War*, http://www.calresco.org/beckerman/nonlindy.htm (accessed 5 April 2011), 1-2.

[8]Headquarters, Department of the Army. ATTP 3-39.20, *Police Intelligence Operations* (Washington, DC: Government Printing Office, 2010), 4-10.

[9]Richard Saferstein, *Criminalistics: An Introduction to Forensic Science*, 9th ed. (Upper Saddle River, NJ: Pearson Education, 2007), 5.

[10]Louis Liebenberg, *The Art of Tracking: The Origin of Science* (Cape Town, South Africa: David Phillip Publishers), v.

[11]Ibid., vi.

[12]Headquarters, Department of the Army, ATTP 3-39.20, *Police Intelligence Operations,* 4-10.

[13]William J. Bodziak, *Footwear Impression Evidence: Detection, Recovery, and Examination*, 2nd ed. (Boca Raton, FL: CRC Press), 2.

[14]Dwayne S. Hilderbrand, "Footwear, The Missed Evidence," http://www.crime-scene-investigator.net/footwear.html (accessed 22 August 2010).

[15]Mark Hansen, "He Tries Men's Soles," *American Bar Association* (May 2010), http://www.abajournal.com/magazine/article/he_tries_mens_soles/ (accessed 17 September 2010).

[16]293 Fed. 1013 (D.C. Cir. 1923).

[17]Saferstein, *Criminalistics: An Introduction to Forensic Science*, 16.

[18]Ibid.

[19]Cornell University Law School, *Federal Rules of Evidence*, Rule 702, http://www.law.cornell.edu/rules/fre/rules.htm#Rule702 (accessed 5 April 2011).

[20]Ibid.

[21]509 U.S. 579 (1993).

[22]Findlaw for Legal Professionals, *Daubert v. Merrill Dow Pharmaceuticals, Inc.*, http://caselaw.lp.findlaw.com/cgi-bin/getcase.pl?court=US&vol=509&invol=579 (accessed 5 April 2011).

[23]526 U.S. 137 (1999).

[24]Cornell University Law School, *Kumho Tire Co., Ltd. v. Carmichael*, http://www.law.cornell.edu/supct/html/97-1709.ZO.html (accessed 5 April 2011).

[25]*Henry v. Ryan*, Dist. Court, D. Arizona 2009, CV 02-656-PHX-SRB.

[26]Findlaw for Legal Professionals, *Daubert v. Merrill Dow Pharmaceuticals, Inc.*

[27]Catherine M. Black, "Legal Implications of the Use of Biometrics as a Tool to Fight the Global War on Terrorism" (Master's Thesis, U.S. Army Command and General Staff College, 2008), 7-19.

[28]Saferstein, *Criminalistics: An Introduction to Forensic Science*, 484-491.

[29]Black, "Legal Implications of the Use of Biometrics as a Tool to Fight the Global War on Terrorism," 18-19.

[30]Saferstein, 486.

[31]U. S. Marshals Service, "Scout Tracking Basic Course" (Certification Course, Fort Richardson and Susitna, AK, May 2000).

[32]Ed Ayyappa, "Normal Human Locomotion, Part 1: Basic Concepts and Terminology," *American Academy of Orthotists & Prosthetists* 9, no. 1 (1997): 10-17, http://www.oandp.org/jpo/library/1997_01_010.asp (accessed 23 October 2009).

[33]Ibid.

[34]The eight adaptive traits is a term commonly used within the field of hoplology as modeled in the original 1998 work of Richard Hayes titled, *Paleolthic Adaptive Traits and the Fighting Man* published by the International Hoplology Society.

[35]Jack Kearney, *Tracking: A Blueprint for Learning How* (El Cajon, CA: Pathways Press), 1.

[36]Ibid., 2.

[37]Liebenberg, 17.

[38]Hardin and Condon, 91, 219, 323.

[39]Ibid., 71.

[40]U.S. Marshals Service, "Scout Tracking Basic Course," May 2000.

[41]Hardin and Condon, 84-85.

[42]Ibid., 228.

[43]Hansen, 3.

[44]Ibid., 4

[45]Michael Hull, "Tracking as Evidence," *Track & Sign* 28.

[46]Ibid.

[47]Southwest Guidebooks, "The Four Corners Fugitive Search-The Largest Manhunt in Western History," http://southwestguidebooks.com/fugitives.htm (accessed 24 July 2010).

[48]Leibenberg, 27.

[49]Ibid., 44.

[50]Bodziak, 2.

[51]Liebenberg, v.

CHAPTER 2

LITERATURE REVIEW

Evidence encompasses a wide array of physical objects, testimony, electronic data, and analyses; it is a key source of police information. Evidence consists of objects, material, or data that can provide proof or a high probability of proof that an incident, association, or pattern will lead to a conclusion or judgment. The thoughts, intuition, and opinions of an analyst or investigator are not evidence; however, they can be critical in forming a conclusion or judgment. Effective evidence collection requires planning, preparation, execution, and training. . . . Evidence collection should be performed as a deliberate and methodical process, unless the tactical situation requires a hasty collection effort. Evidence should also be handled by as few personnel as possible to avoid contamination and risk of breaking the legal chain of custody.

— Headquarters, Department of the Army
ATTP 3-39.20, *Police Intelligence Operations*

Overview

This study will cover behavior analysis, gait analysis, and spoor analysis, which is contained in the BGS paradigm. No single paper can be found addressing all three subjects simultaneously. Furthermore, this research involved extensive studies to construct the theoretical framework for this study. The literature review covers the most significant contributions to the thesis work. The order of appearance of the previous research reflects the way in, which the researcher has approached the research questions in reference to BGS paradigm and OCISE algorithm construction. The review of research in the three areas above exposes one large gap: many researchers have studied various parts of behavior, gait, and spoor separately; however, no researcher has studied them together addressing why spoor evidence is influenced by human behavior from a forensic spoorological perspective; which includes baseline and human behavior identification.

29

Behavior Analysis

It is common knowledge as to what social behavior is. The sophisticated action that is directed toward other people.[1] Therefore, behavior is psychomotor and locomotor action. This we know as either good behavior or bad. It is the tolerability of human behavior assessed comparative to social norms and controlled by numerous means of social governance.[2] Behavior in reference to spoor evidence must be visually observable to be studied spoorologically. In practice, "publicly" observable body-bound and action-bound processes (gait, movement, etc.) and "privately" non-observable brain-bound processes (thinking, feeling, etc.) are interconnected and comprehensive in both micro and macro human capabilities.[3]

In Thomas-Cottingham's book, *Psychology Made Easy* descriptions of the fundamentals of all disciplines of psychological theories are articulated. The physiology of sensation, perception, cognition, motivation, and emotion are referenced as the biological (nature) side of all humans.[4] In addition, the environmental influence of our experience (nurture) is advocated. The debate of nature versus nature is proffered and most psychologists agree that we as human beings are a product of both our genetics (brain-bound) and our conditioning (body-bound and action-bound).[5]

Huitt's research of human behavior on *A Systems Approach to the Study of Human Behavior* exemplifies the systems approach to the study. A framework of the systems approach is linked to the three major aspects of human beings (mind, body, spirit). This systems approach is interrelated and cannot function without all parts. The mind is composed of cognition (knowing, understanding, thinking), affection (feelings, emotions, attitudes, predispositions), and conation (volition, will, intentions to act, reasons for taking action). This framework conceptually delineates that the mind receives information through the five senses and manifests action through the body. The body is composed of genetic influences, bodily functioning, and overt behavior. The framework takes into consideration what Huitt calls a feedback loop between overt behavior (responses) and resulting stimuli from the environment. Lastly, Huitt's framework recognizes the interrelation of biological influences and spiritual influences on human development and functioning.[6]

In *Evolve your Brain: The Science of Changing Your Mind*, Joe Dispenza, D.C. suggests that all human thoughts set off biochemical reactions that lead to behavior.[7] Through repetition are unconscious thoughts produce automatic acquired patterns of behavior. Therefore, they are almost involuntary in nature and build behavioral patterns that are hard-wired neurologically in the brain by memory. We have memories as automatic programs that assist in a person being able to thrive in the world. As we process the same thoughts day in and day out, we become neurosnaptically wired equal to our past experiences in our environment. These neural networks are hard-wired from our repeated "thoughts, actions, behaviors, feelings, emotions, skills, and conditioned experiences."[8] This is important because it allows the body ability to respond quickly to stimuli in the environment. The stimuli from the environment is processed through an external cue by one of the senses (usually sight). Then a thought, which is stimulated by the external cue turns the brain into a processor for retrieving an associative pattern stored in the brain. Once retrieved, the associative is run as an automatic thought program or stream of consciousness, which prompts the body to move to accommodate the thought from the brain.[9]

31

Using Dispenza's research, the following is the process for activating normal walking behavior to start traveling from point A (standing on the ground next to a rock outcropping surveilling a house) to arrive at point B (the window of the house to look inside):

1. Your thought of moving from your current position to end state position creates the first series of action potentials in your brain.

2. Your eyes see the window and initiates the second series of action potentials.

3. Your occipital lobe (the part of the brain responsible for vision) registers the image of what you see.

4. The temporal lobe (responsible for association in conjunction with memory storage and learning) associates the image of what you see with what it remembers of windows, which then creates another series of action potentials.

5. The frontal lobe (responsible for higher mental activities) allows you to maintain your attention while you intentionally begin moving toward the window.

6. When you begin to formulate and integrate the movement walking toward the window, the frontal lobe and pariental lobe (the motor portion of the brain, also responsible for language mechanisms and general sensory functions) helps you initiate the action movement in your legs, feet, and HAT, and triggers your sensory anticipation of what walking will feel like.

7. The pariental lobe allows you to feel that you are walking toward the window- you can sense your body moving through the gait cycle, each foot as it moves

and is placed on the ground, the transfer of weight to each foot during locomotion.

8. At the same time, the cerebellum (responsible for coordinating voluntary muscular activity) directs the body's fine motor movements to adjust to the motor movements to maintain proper balance. Without the cerebellum, you might take a step with the foot but end up placing it outside your range of balance thus falling to the ground.[10]

Another aspect of Dispenza's research that is of worth for this thesis is his indepth study into the survival part of the brain (The midbrain is responsible for responding based on our needs to survive in the environment), which is much different from the reasoning part of the brain (The forebrain is responsible for responding based on thinking). This is important because the body shuts the forebrain down when under stress and implements the autonomic nervous system to trigger a fight or flight response to insure survivability. Many of the stresses of life while moving within an environment transfer through the gait as behaviors inscribed on the ground as spoor evidence.[11]

Dispenza clearly shows that the human body does not display any behavior without the mind (conscious or unconscious thought) directing it. All body-set and action-set movements start in the mind (the brain in action) with a single thought. This research is crucial and lays a salient foundation to understanding the BGS paradigm in reference to behaviors influence on gait and spoor evidence through linearity.

Gait Analysis

Ed Ayyappa's study of normal human locomotion describes the basic principles of normal walking and provides a common language of human movement.[12] This study

33

lends credibility to the study of spoor evidence because an understanding of gait is the window to know human locomotion, the foot platform that accepts locomotion and provides structural stability, and the thoughts that project the causes of human movement.[13] Kale et al.,[14] Thorton, Pinto, Shiffrar,[15] and Lee,[16] all studied human gait for the purpose of recognizing people by their gait from a distance visually.

Gait itself, from a recognition perspective, reveals that humans have the ability to recognize people from watching gait visually, which indicates "the presence of identity information in the gait signature."[17] These studies help to recognize that gait movements, if witnessed by direct visual observation, isolates gait as unique to one person over another therefore articulation of human gait discrimination is verifiable.[18] This uniqueness attributed to gait in people is also then unique in how gait signatures are transferred into the substrate through locomotion. Since gait has unique properties of recognition, then the GFS are also unique to the gait that produces the signature.[19]

In Nowlan's study through the use of accelerometer gyro forces, an accuracy rate of 95 percent is achieved in gait identification to one person only. This accuracy is complimentary in analysis of gait cycle individuality observed in the other papers on human gait. The gait individuality conclusions in Nowlan's work although accurate were instrument driven. Human accuracy in observation of spoor evidence as to individual gait is new and significant data is unavailable to this study and therefore warrants the research.[20]

Spoor Analysis

Many books describe the use of sight as an asset in gathering information from the ground. Many people can look into an area of surface soil and vegetation. The process of directing your gaze to an area does not equate to understanding what is in the area.[21] It is important for the tracker to know why things are seen or what must be looked for in any scene. There are five factors that draw attention to the eyes and determine what can actually be seen.[22]

Shape

There are shapes all around us. All the shapes that draw our attention are usually manmade or are associated with humans. The most distinct is the human body itself. These shapes stand out to the human eye. All humans notice these common shapes daily, so their eyes are naturally trained to see them. Also, noticeable to the human eye are the things worn or used, i.e. hats, coats, boots, tools, guns, etc. All these are even more noticeable when standing or sitting against a background that is a distinct contrast to the object. This is what makes an obvious track noticeable to the untrained eye and sign noticeable to the trained eye.

Robert Speiden in his book, *Foundations for Awareness, Signcutting and Tracking,* describes shape and its related modalities this way:

> The shape of a track is the overall area as defined by its edges. . . . The edge of a track or sign is the boundary between the disturbance and the baseline. . . . If the edge of the track is sharp be a fairly recent track, especially in a sandy soil. A clay-based soil can also hold an edge for quite a while before deteriorating. The edge of the track will generally wear down and round off as weather and other aging forces act on it.[23]

It is rare to see a complete print with all the details while on a crime scene because of the weathering process and other external factors as in contamination by other tracks, vehicles, animals, etc. Even though external forces are a reality to the tracker, there are still numerous pieces, fragments, and remnants of track impressions due to many external factors. The human eye must not become lax in pursuit of only something perceived as obviously clear (as in soft ground or snow). Once the eyes are trained, even the smallest visual spoor clue will become a beacon or billboard.

Ab Taylor and Donald Cooper, *Fundamentals of Mantracking: The Step-By-Step Method,* explain "If a tracker looks for certain signals or visual cues (cue: a stimulus that guides behavior) that catch the eye, rather than tracks or prints, then, in the end, far more will be seen."[24]

Shadow

Shadowing is cast by a light source. The most common is the natural daily sunlight. Other forms of light include the moonlight and artificial light; like a flashlight. When a shadow is cast this is a signal to the eye that spoor is present, something is built-up and is blocking the clear light. This forms a shadow. One point with shadowing, shadows move based on the position of the light source. When the sun moves during the day, the natural shadowing moves as well. This, at times, can hide spoor within the shadows. Until the light source is favorable again to the spoor, the tracker will be blinded to the spoor.

Roland Robbins, *Mantracking: Introduction to the Step-By-Step Method,* asserts:

One of the most important elements in tracking is light. Not only is its quality important, but also the direction from, which it is coming and the angle at, which it strikes the surface you are looking at. . . . Under such conditions, even small irregularities or bumps will cast long shadows, and therefore provide maximum contrast with the surrounding surface. The ideal light angle is in the early morning and late afternoon when the sun's rays are nearly parallel to the earth's surface. To take maximum advantage of this leadel light angle you must position yourself so that the area you are searching lies between you and the light source.[25]

The sun and moon, as a light source, cannot be controlled and therefore requires the tracker to change positions and angles of vision in order to enhance the light to an advantage. Artificial light, however, is another matter. It can be controlled, which allows the tracker to manipulate the angles to best suit the eyes.[26]

When a foot falls in the surface soil it leaves an impression. The foot creates with gravity its own contours or relief. Soft soil produces bigger contours than does harder soil. This relief in the soil interacts with the light source to produce a shadowing effect. The eye, when trained, picks up the shadowing that the light source casts because of the relief that the foot has left.

Silhouette

Silhouetting is the very nature of outline and shape. The best example of silhouetting is when the eye sees something dark against a light background or something light against a dark background. Many times while at the crime scene the tracker will be looking for this eye catching principle.

Skylining is familiar to most people and is related to the concept of the silhouette. It is particularly familiar to those who have had military training. When a track is placed in soil by the foot, the contouring and relief also cause this silhouetting principle. The new shape contained in the soil will cast or skyline against the surrounding soil and vegetation. This is known as the spoor baseline, which is different from the baseline contained within the OCISE algorithm. Yet, understanding the baseline of any environment is crucial to what is normal, which then accentuates what is an anomaly. Kevin Reeve, Chief Instructor and Director of Onpoint Tactical Tracking Services, in a personal treatment to tracking described the baseline as a symphony when he wrote:

The baseline symphony is the sound of the woods at rest. It is the normal, background noise and activity level of the forest (or any other environment). It must be established in the mind of the scout at any time of day or night so as to provide a baseline for comparison. Once he knew that the baseline symphony of the forest is a certain level at a certain time of day, then any variation from that baseline must be created by something out of the ordinary.[27]

Shine

This represents color and texture to the eyes.[28] The medium is the surface soil and vegetation. When anything contrasts on the soil surface and vegetation, it should attract the eyes. Many things at a crime scene will catch the eye. Shine is a concept that the eye will collect much of. Shine can be expressed in understandable terms by equating it with the shine that is present when driving down the road in a vehicle and the sun reflects off the hood. It, in essence, hurts to look at it. This is what happens when light reflects off anything. While at a crime scene, the light source will be used to bring shine to the eyes. Sometimes this happens naturally and sometimes it happens after much practice of controlling the light source or the right position in relation the light source.[29]

During normal walking gait, the foot and body leave things behind. An example would be the spoor left through a grassy field. The grass stands in natural patterns to its interactions with the weather. When the quarry walks through such a field they leave their gravity impressions upon the soil and grassy vegetation. The light source will shine contrast off of both the grass that is unaffected and the grass and soil that is affected by human passage. There will be a distinct contrast and the reflection on the light source will bring this to the trackers eyes.

Spacing

The spacing of objects in regular patterns is contained within the BGS paradigm and suggests a human interaction or baseline. Nature, however, places things randomly or nonlinear. Spacing is represented best by human movement patterns of stride and step intervals during locomotion that are registered into the substrate by footfalls of a trackmaker. When spoor knowledge of what to look for in the quarry's baseline, the quarry will leave information in regular intervals to be seen by the tracker.

Focusing the Mind's Eye

The three levels of spoor analysis as used in each fundamental point of observation to help the tracker focus the mind on spoor evidence are primary, secondary, and tertiary. Primary analysis is necessary on its own merit as well as in confirming the analysis that is observed in the secondary and tertiary areas.[30]

Primary

Primary analysis is looking for spoor that is conclusive evidence of the passage of the quarry. Conclusive evidence indicates on its own merit that the quarry moved through the area. This means to the tracker that they are still in maintenance with the spoor story. Some examples of primary spoor are a clear footprint and identifiable litter.

Secondary

Secondary analysis is looking at spoor that is substantiating evidence of the passage of the quarry. Substantiating evidence indicates to the tracker that they need to collect many samples in order to draw the correct inferences. Secondary spoor may be the quarry moving through an area and it may not. The tracker through the application of "continuity" and "constancy" contained in additivity will be able to connect each piece of secondary spoor and by stringing this evidence together, come to concrete extrapolations about the quarry behaviors. Some examples of secondary spoor are rocks out of place, bruised vegetation, broken twigs, and others.

Tertiary

Tertiary analysis, also called extended, is calculating all other gathered evidence related or not related to spoor evidence. Some examples of this type of spoor are animals being flushed out, interviews with witnesses, etc. Contamination of the spoor is also placed within this classification. This could be the spoor of other investigators, technician and scientists as they move about in or near the crime scene, or it could be an animal's spoor. Any sign that is not the quarry's spoor would be classified as tertiary spoor.

Zones and Characteristics

David Scott-Donelan, *Tactical Tracking Operations: The Essential Guide for Military and Police Trackers* explains the zones and characteristics connected with the leaving of spoor evidence along the trail as the body employs the locomotor apparatus during locomotion.[31]

Ground Spoor

This spoor is also known as bottom spoor. This zone is any spoor found below the ankles. All three classifications of spoor analysis above would fall into this zone. If the tracker is at a crime scene and comes across a place where a moose crossed over human spoor stepping on the prints of the quarry, this is spoor within the zone of ground spoor. It is also contaminating your spoor evidence, which would be also classified as tertiary spoor. This type of contamination is great to come across because it helps with the process of evaluating age thus assisting the tracker with an accurate track erosion computation. Ground spoor is footprint; ground and twig scuffs; any transfer or displacement of dirt, water, or leaves; rocks dislodged or compressed; animal scat; human feces; and others.

Aerial Spoor

This zone is any spoor found above the ankles. All three classifications above can be found in this zone; however, secondary spoor and tertiary spoor are more prevalent. When a quarry walks in the woods off a manmade trail, they must move around and through vegetation. This causes damage to the brush they maneuver to maintain proportionality. As the tracker builds a mental picture for the approximate size of the quarry, they begin to look to that height for this kind of spoor and sign. Examples of this spoor are scrapping, cutting, and bending of vegetation; hand holds or antler scrapes on branches and against trees; morning dew dislodged from vegetation; torn clothing hanging from thorns and branches; cobwebs broken; shine on vegetation pushed forward in direction of movement; etc.

Conclusive and Substantiating Evidence

Both Jack Kearney[32] and David Scott-Donelan[33] designated one or more of the four spoor evidences for maintenance of linearity within the BGS paradigm.

Regularity

According to Scott-Donelan, "Regularity is the eye-catching effect of unnatural, man-made patterns."[34] Regularity is shapes that are never to be seen in nature unless a human brings it there. If straight lines, circles, stars, and square marks are found in nature in calculated formations, this was made by the human animal. The shoes of man are made in such formations. They are unique to man only. Nature leaves no such shapes.

Flattening

When the human foot strikes the surface soil during locomotion it compresses the area stepped on so that there is a flat spot. There are animals that also leave flat spots. This foot pressure flattens anything underneath the applied pressure such as dirt, twigs, rocks, leaves, etc. This is another characteristic that is found from the application of human shoes to any substrate.

Color Change

Mostly color change occurs from a context of moisture and/or temperature differences. For example, when a rock is kicked loose a dark spot of dirt surrounded by lighter soil or the dark underside of the rock against the lighter soil will be seen. Either effect from the foot leaves a contrast that the eye picks up. While the rock sits in place, undisturbed, it shades the underside from exposure to heat derived from the sun. The topside of the rock has been heated day after day causing a drying out. The under belly of the rock has the combination of heat produced from the topside, which penetrates the rock, indirectly, heating the cool unexposed earth beneath and the soil moisture factor, which comes up from the water table. This principle also applies to overturned sticks, leaves, dirt, litter, etc. When color change is notice in the continuity of a spoor-chain signature, then, it was caused by a quarry.

Disturbances

All classifications and characteristics of spoor can be described as a disturbance.[35] If a quarry walks along a path and rubs an exposed root with a lugged boot, this piece of evidence is a disturbance. It also can be ground spoor or aerial spoor depending on, which zone it is found in. It can be primary, secondary, or tertiary spoor depending on who left the evidence and where it was found. Depending how deep the root is cut, it could be permanent or temporary spoor (addressed next). All spoor evidence is a disturbance.

Two Levels of Registration

Bob Carss, *The SAS Guide to Tracking*, characterizes the two levels of registration.[36]

Temporary

This spoor evidence will go away with the passage of time. It is affected by the weathering process. When a footfall strikes the surface soil it leaves an impression. Through wind, rain, heat, and cold, the formation that was left will start to erode and return to a natural contours. When you look at the ground spoor, the spoor will be at varying stages of age. The more identifiable the stages become through visual experience, the more the spoor evidence will indicate how long ago the quarry passed through the crime scene. When looking at aerial spoor, such as broken cobwebs, the spider that made it will begin the repairs immediately. Much time should be spent in logging the effects of time on spoor evidence through direct visual observation.

43

Permanent

Of course, this level of registration implies perceptually that this spoor will never deteriorate. Although, all spoor evidence will eventually return to as close as possible to its original state, this spoor evidence will not go away, easily. It stays around for a long while. Nothing is forever, when it comes to spoor evidence. Permanent spoor recovers slowly. So slowly, it seems that it is permanent. When coming across a birds' nest one day, for example, it will still be visible many years later. The manmade plastic invention can be preserved for 40 or more years. Human litter or discardables will be present long after they are observed. The most noticeable permanent spoor is when vegetation is cut. This cutting of a tree, for instance, will slowly start a scarring action that will usually be permanent in one form or another. The tracker, through many hours of direct visual observation of such vegetation damage and the healing process it undergoes can with much accuracy, determine when it was cut.

Other General Spoor Analysis Topics

There is a substantial amount of information available on tracking; some of, which focuses on the backgrounds, careers, and achievements of specific trackers. These articles and books support tracking because they explain how to use tracking within various mission parameters (see figure 4).[37] The literature also gives examples of actual tracking cases. For instance, David Sacks article explains that drug smugglers who come across the Mexican border leave deeper tracks because of the heavy load they are carrying and that tracking is used on surfaces ranging from linoleum to forest floors by looking at sign [spoor] such as footprints, broken twigs, bent blades of grass and scuff marks.[38] Dwayne Hildebrand's article discusses how to preserve and collect footwear evidence. More specifically, it explains how to search a crime scene for footprints and how to photograph and cast them.[39]

The professional articles also deal with footprint evidence and tracking. These articles contain more detailed information about the benefits of tracking and gathering footprint evidence. Tom Hanratty's article combines clear examples of police use of tracking with explanations of what can be learned about the suspects from their tracks.[40] Hanratty uses these examples to illustrate why it is important to recognize and preserve footprint evidence at a crime scene.

Almost all of the other literature includes information on how to lift impressions, such as by casting, which is of minor concern to the science of spoorology. The other problem inherent with these books and articles explaining footprint lifting techniques is that it focuses a police officer, investigator, and crime scene technician on only the spoor evidence that is infrequent--obvious footprints, when in reality, there is much more spoor evidence present at the scene. Based on this type of instruction, for example, if an investigator tries to determine whether a subject was running, walking, or carrying a heavy object they will only be able to do so when there are many clear prints, such as in mud or snow. If an investigator was trained in tracking they could locate the less obvious spoor and gather more information about the subject.[41]

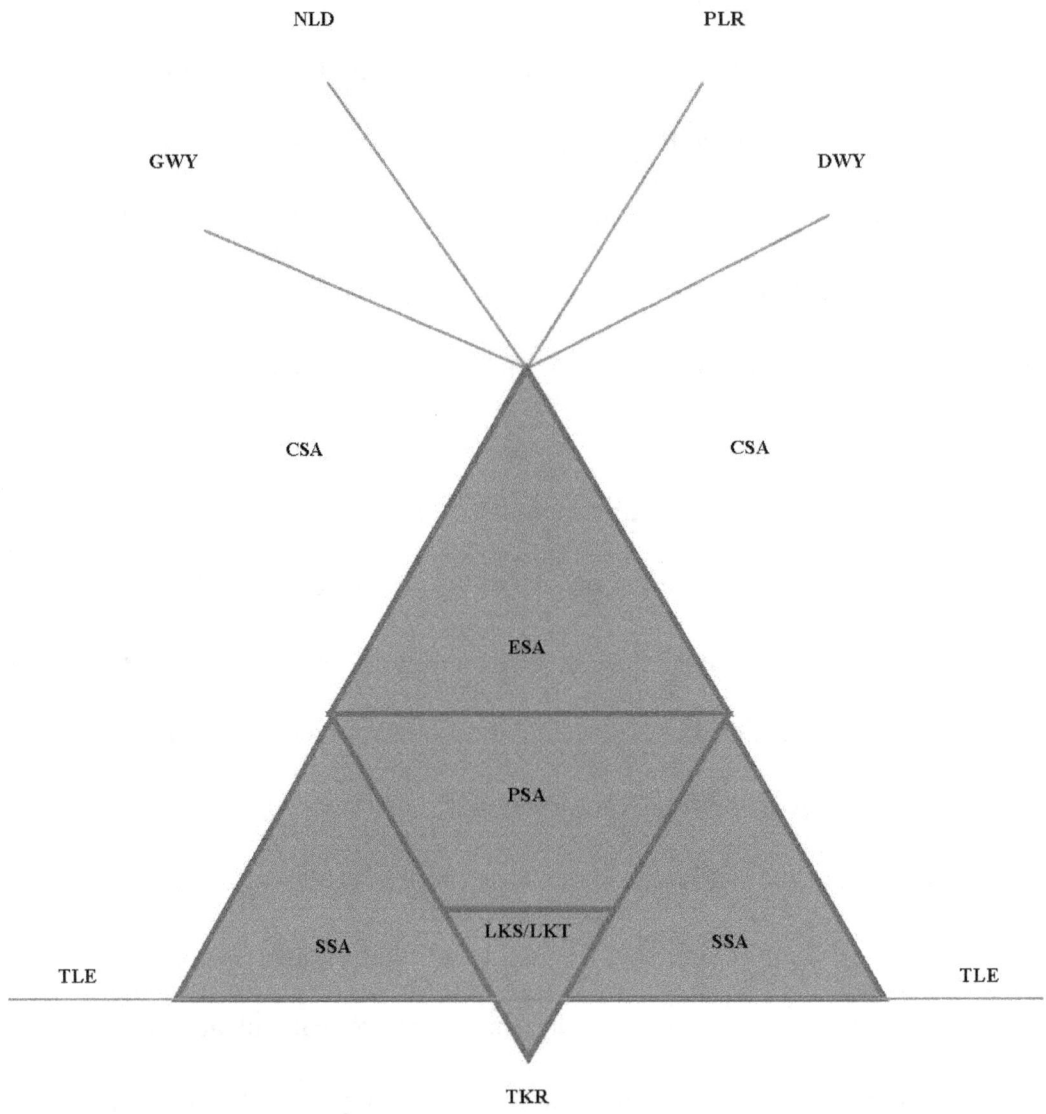

Figure 4.   Tracker's Triangle

Source: Created by author.

Note: The tracker's triangle is a mental model, which the tracker should use to manage a quarry's spoor. It focuses a tracker's visual search pattern on locating the next spoor in the gait-footfall sequence thus mitigating extensive lost trail procedures while facilitating continuity of spoor and thus human behavior discovery.

These books and articles provide limited instruction for dealing with footprint evidence, but they identify some important uses that it can provide. Paul Vallandigham, in a tracking paper presented at the annual symposium of the International Society of

Professional Trackers lists 22 things that footprints tell the astute tracker. Some of these

are:

1. How many people were involved.
2. Where the people went at the scene.
3. What they could have touched (for fingerprints).
4. Where they went when they left the scene
5. The sex of the person
6. Whether they were right handed or left handed
7. How tall they were.
8. Approximate weight.
9. When they were there (time of the crime).
10. Help with a positive identification of the suspect (s) from stride length, step interval, pitch angles, straddle, trail width, and personal identifying pressure releases.[42]

Another important part of a criminal investigation is protecting and preserving the

crime scene. The crime scene is where one gets most of the physical evidence and

because of its importance in solving crimes; it needs to be protected from destruction.[43]

All of the texts and articles mentioned above also highlight the importance of protecting

the crime scene. They explain how evidence can easily be destroyed and that a systematic

search may be performed too quickly, causing areas of peripheral evidence to be

overlooked or minimally examined. Overall, these books adequately recognize the

importance of protecting the crime scene. However, it is equally important to know what

to protect. An investigator needs to be aware of the spoor evidence that is present at a

crime scene so they do not lose important evidence pertinent to the investigation. For

example, if a homicide victim is found outdoors, an investigator with tracker training

would know to secure a large enough area to protect spoor evidence that is not right

around the body. The basic information in this literature is not enough to help officers, investigators, and crime scene technicians to save the evidence that is frequently

destroyed or to teach them how to use spoor evidence to further an investigation.

Legal Precedence Case Review for
Spoor Evidence Admissibility

Chapter 1 discussed the history of scientific, technical, and specialized knowledge

for admissibility in court. The U.S. Marshals Tactical Tracking Unit manual, *Scout

Tracking Operations Course: Law Enforcement-Certified Tracker I, II, III*, contains a

more thorough list of legal precedence cases.[44] A review of the cases was conducted and

the salient ones are given by topic here. The terms footprint, shoeprint, footwear

impression, shoe imprint, etc. are all common terms understood by everyone. In reference

to the above terms as descriptions of what the eyes see imprinted in the various crime

scene substrates by officers, investigators, and crime scene technicians, the science of

spoorology uses the term spoor.

Shoe Print Evidence

The general principle that footwear impression evidence is reliable technical or

other specialized knowledge under *Daubert* and *Kuhmo Tire* and is generally admissible

expert testimony in a criminal case.[45] The correspondence of shoeprints, found in

connection with a crime, in regards to the track of a shoe of an accused is admissible to

identify the guilty party.[46] Evidence regarding measurements taken off footwear

impressions found at a crime scene were the same as measurements taken off the

defendant's shoes was admissible. Defendant's objections to the evidence were "not well

taken, and rather applied only to the weight of that evidence."[47]

Competent Evidence

Footprints found at the scene of a crime has long been held admissible as competent evidence in an attempt to identify the accused as the guilty person.[48]

Shoeprints Alone Sufficient to Convict

Shoeprint evidence, standing alone may be sufficient to sustain a conviction. Where there are sufficient general and individual characteristics, that would provide a basis for a positive identification, footwear impression evidence may be as reliable and trustworthy as any other evidence.[49]

Shoeprints Sufficient to Sustain a Conviction

Footprints leading to and from the assault victim's house, together with other evidence, was sufficient to sustain a conviction.[50]

Probably Made Conclusions

An expert in footwear identification testified that a tennis shoe found in the defendant's closet probably made the shoe print found outside the victim's home, stating all the facts he observed in comparing the tennis shoe and the footprint. The court held the result of "probably" is admissible, even though it is not a positive identification. The lack of exactness in the opinion went to the weight the jury could give the evidence, and not to its admissibility.[51]

The Required Number of Points of Identification

Higher court upholds findings of trial court that the six similar individual characteristics could all be the result of coincidence. Were there only one similar characteristic, we would be more inclined to accept this argument. However, we believe that even one individual characteristic, depending on the nature and uniqueness, could be enough for a valid comparison.[52] Footprint evidence is admissible, if tracings have sufficient individual characteristics to make the comparison reliable.[53] Two identifying marks were held to be sufficient to establish positive proof that the impressions at the crime scene were left by the boots the defendant was wearing at the time of his arrest.[54]

Establishing Time of Impression

Describing how the impression was found, the court said it could be reasonable inferred that the impression had been made after a heavy rain, which occurred the afternoon before the crime. The impression tended to show the defendant had been in the vicinity of the crime scene near the time of the commission of the crime.[55] Permitting a police officer to describe footprints in snow as "fresh" was not error.[56]

Time of Placement Requirements

Evidence of shoeprints has no legitimate or logical tendency to identify an accused as the perpetrator of a crime unless the attendant circumstances support this triple inference: (1) that the shoeprints were found at or near the place of the crime; (2) that the shoeprints were made at the time of the crime; and (3) that the shoeprints corresponded to shoes worn by the accused at the time of the crime.[57]

Lay Witness/Non-Expert Testimony

Testimony of a non-expert police officer concerning similarity of test impressions made with the defendant's shoes and footprints found at the scene of a burglary was admissible.[58] A non-expert was properly permitted to testify that "footprints" had been made by tennis shoes. The non-expert was properly permitted to testify that the design of the impression was the same pattern at the sole of the shoe worn by the defendant.[59]

Officer Observed Shoeprints/Shoes

Trail judge did not error in admitting testimony of Police Officer concerning visual comparison he made of wavy pattern in shoe print found outside basement window at scene of burglary with pattern on tennis shoe taken from defendant after arrest.[60] Testimony of a police officer concerning a visual comparison he made of a footwear impression, which he observed at the crime scene with the pattern he observed on the defendant's shoes after the defendant was arrested was not admitted in error. Even if the officer was not an expert, his observation was a fact issue open to the sense of sight.[61]

Warrantless Arrest

A distinctive set of bootprints leading from the defendant's car to a recently burglarized home, finding the defendant one hour after the crime approximately one mile from the crime scene, wearing boots matching the distinctive boot prints provided sufficient grounds for a warrantless arrest.[62] Footprints in snow, leading from the crime scene to the place where the defendant was located, consistency between the footprints and the type of boots worn by the defendant, and other factors provided probable cause to arrest the defendant.[63]

Bill of Rights-Amendments

Grand jury's directive that a suspect submit his feet and shoes for ink printing did not constitute a "search" of his person subject to the Fourth Amendment. The government needs only make a preliminary showing that the prints are at least relevant to an investigation being conducted by the grand jury and properly within its jurisdiction, and are not sought primarily for another purpose when it invokes the district court's power to enforce obedience with such a subpoena.[64] Boots may be taken from a suspect as an incident to a lawful arrest for investigation of a crime and the accused's connection with the commission of that crime. This is not testimonial evidence and does not come within the Fifth Amendment's protection against self-incrimination.[65] Footwear impressions found where the vehicle was stolen were identical to footwear impressions leading from where the vehicle was recovered to the defendant's house, and the impressions bore a distinguishing imprint. The defendant had no reasonable expectation of privacy in the soles of his shoes.[66]

Summary

This chapter discussed the numerous literature regarding the component parts (behavior, gait, and spoor), which make up the forensic spoorology BGS Paradigm. The next chapter presents the OCISE algorithm, as criteria methodology to assess the extent to, which spoor evidence will elucidate human behavior thus addressing the primary and secondary research questions.

---

[1]W. Huitt, "A Systems Approach to the Study of Human Behavior," *Educational Psychology Interactive*, http://www.edpsycinteractive.org/materials/sysmdlo.html (accessed 16 September 2010).

[2]Alison Thomas-Cottingham, *Psychology Made Easy* (New York: Broadway Books), 28.

[3]Huitt, 2.

[4]Thomas-Cottingham, 1-2.

[5]Ibid., 6-7.

[6]Huitt, 1-8.

[7]Joe Dispenza, *Evolve Your Brain: The Science of Changing Your Mind* (Deerfield Beach, FL: Health Communications), 2, 3, 43.

[8]Ibid., 240.

[9]Ibid.

[10]Ibid., 85.

[11]Ibid., 111.

[12]Ayyappa, *Normal Human Locomotion, Part 1: Basic Concepts and Terminology.*

[13]Ibid.

[14]A. Kale et al., "Gait Analysis for Human Identification," http://www.cfar.umd.edu/~cuntoor/kale03gait.pdf (accessed 19 November 2010).

[15]Ian M. Thorton, Jeannine Pinto, and Maggie Shiffrar, "The Visual Perception of Human Locomotion," *Cognitive Neuropsychology* 15, 6/7/8 (1998): 535-552, http://psychology.rutgers.edu/~mag/reprint_pdfs/TPS98.pdf (accessed 19 September 2010).

[16]Lily Lee, *Gait Dynamics for Recognition and Classification* (Cambridge: Massachusetts Institute of Technology, 2001).

[17]Kale et al., 1.

[18]Ibid.

[19]Lee, 3.

[20]Michael F. Nowlan, "Human Identification via Gait Identification Using Accelerometer Gyro Forces," http://cs-www.cs.yale.edu/homes/mfn3/pub/mfn_gait_id.pdf (accessed 13 July 2010).

[21]T. J. Cunningham, *Scout Tracking Operations: Basic Trailing & Surveillance Concepts for Operational Trackers* (Cheyenne, WY: Lost Trail Institute, 2004), 79.

[22]Adrian Gilbert, *Sniper* (New York: St. Martin's Press, 1994), 206.

[23]Robert Speiden, *Foundations for Awareness, Signcutting and Tracking,* (Christiansburg, VA: Natural Awareness Tracking School, 2009), 142.

[24]Albert "Ab" Taylor and Donald C. Cooper, *Fundamentals of Mantracking: The Step-By-Step Method,* 2nd ed. (Olympia, WA: ERI International, 1990), 38.

[25]Roland Robbins, *Mantracking: Introduction to the Step-By-Step Method* (Montrose, CA: Search and Rescue Magazine, 1977), 38-39.

[26]Cunningham, 80.

[27]Kevin Reeve, "Tracking," On Point Tactical Tracking Services.

[28]Speiden, 140, 142.

[29]Robbins, 39.

[30]Cunningham, 87-88.

[31]David Scott-Donelan, *Tactical Tracking Operations: The Essential Guide for Military and Police Trackers* (Boulder, CO: Paladin Press, 1998), 30-31.

[32]Jack Kearney, *Tracking: A Blueprint for Learning How* (El Cajon, CA: Pathways Press), 55-56.

[33]Scott-Donelan, 30-31.

[34]Ibid.

[35]Bob Carss, *The SAS Guide to Tracking* (New York: Lyons Press, 2000), 25.

[36]Ibid.

[37]Cunningham, 2004.

[38]David Sacks, "Tracking," *Gazette* 61 (February/March 1999): 10-12.

[39]Hilderbrand.

[40]Tom Hanratty, "Sherlock Holmes, Master Tracker, Part 1." *Track & Sign* 29: 3-5.

[41]Hardin and Condon, 224.

[42]Paul Vallandigham, "Tracking" (Paper presented at the Annual Symposium of the International Society of Professional Trackers, Petaluma, CA, 22-24 October 1999).

[43]Saferstein, 39.

[44]U.S. Marshals Service, *Scout Tracking Operations Course: Law Enforcement-Certified Tracker I, II, III* (Anchorage: District of Alaska Tactical Tracking Unit, 2000), 54-65.

[45]*United States v Allen*, F. Supp. 2d, 200 WL 1299779 (N.D. Ind. June, 2002).

[46]*McClard vs. U.S.* (CA 8 ARK.) 386 F. 2d. 495 (1967).

[47]*Patterson vs. U.S.*, 62 F.2d 968 (10th Cir. 1963).

[48]*People vs. Kozlowski*, 95 ILL. App. 2d, 464 (1968).

[49]*People vs. Campbell*, 146 ILL.2d. 363, 586 N.E.2d 1261 (1992).

[50]*Smith vs. Commonwealth*, KY. 375 S.W. 2d 819.

[51]*State vs. Kelly*, 111 ARIZ. 181, 526 P.2d 720 (1974).

[52]*People vs. Campbell*, 134 ILL. 2d. R. 315(a) (1991).

[53]*State vs. Hall*, 286 MINN. 424, 176 N.W.2d 254 (1970).

[54]*Sheppard vs. State*, 239 ARK. 785, 394 S.W.2d 624 (1965).

[55]*Ferrell vs. Commonwealth*, 177 VA. 861, 14 S.E.2d. 293 (1941).

[56]*Commonwealth vs. LePage*, 352 MASS. 403, 226 N.E.2d 200 (1967).

[57]*State vs. Palmer*, 230 N.C. 205, 52 S.E.2d 908 (1949).

[58]*People vs. Lomas*, 92 ILL. App.3d 957, 416 N.E.2d 408 (1981).

[59]*State vs. Hairston*, 60 OHIO App.2d 220, 14 O.O.3d 191, 396 N.W.2d 220 (1977).

[60]*State vs. Gardner*, MO. App. 700, S.W. 2d. 172.

[61]*State vs. Gardner*, 700 S.W.2d 172 (MO. App. 1985).

[62]*State vs. Schofield*, 114 N.H. 454, 322 A.2d 603 (1974).

[63]*State vs. Birmingham*, 122 N.H. 1169, 453 A.2d 1329 (1982).

[64]*U.S. vs. Ferri*, 778 F.2d 985 (3rd Cir. 1975) cert. denied 476 U.S. 1172.

[65]*McClard vs. U.S.*, 386 F.2d 495 (8th Cir. 1968).

[66]*Downey vs. U.S.*, 263 F.2d 552 (10th Cir. 1959).

CHAPTER 3

RESEARCH METHODOLOGY

Whenever a person walks through an area, whether it be at home or in the wilderness, evidence is left of that passage. A person must contact his or her environment in order to travel by foot. In fact, walking, the most common type of unaided travel, requires a person to come into contact with their environment approximately once every 18 to 20 inches. Some disturbance (sign) is made through that contact and the first phase of tracking incorporates detecting this sign.[1]

— Albert "Ab" Taylor and Donald C. Cooper,
Fundamentals of Mantracking: The Step-By-Step Method

Overview

The last chapter discussed the body of research regarding the component parts

(human behavior, gait, and spoor), which make up the forensic spoorological BGS

Paradigm. A voluminous amount of research concerning the spoorological parts is

available; however, research concerning the nesting sequencing of the parts to one whole

is non-existent. Since the purpose of this research is to determine if and to what extent

human behavior can be gleaned from the spoorological record, it is necessary to develop

a methodology to analyze various gait-footfall sequences that allow for the use of OCISE

to identify one person from another to include the behaviors attached to the spoor-chain

signature left at the scene. This chapter presents the OCISE algorithm within a Uniform

Scale Format methodology to assess the extent to, which spoor evidence will elucidate

human behavior thus addressing the primary and secondary research questions.

This researcher will confer with the thesis committee to ensure that the projected

methodologies comply with the scope of this thesis before proceeding. With the

methodologies approved, it is necessary to follow the best practices of the scientific

57

method in order to address viability and reliability necessary for the court admissibility of spoor evidence as shown in the previous two chapters.

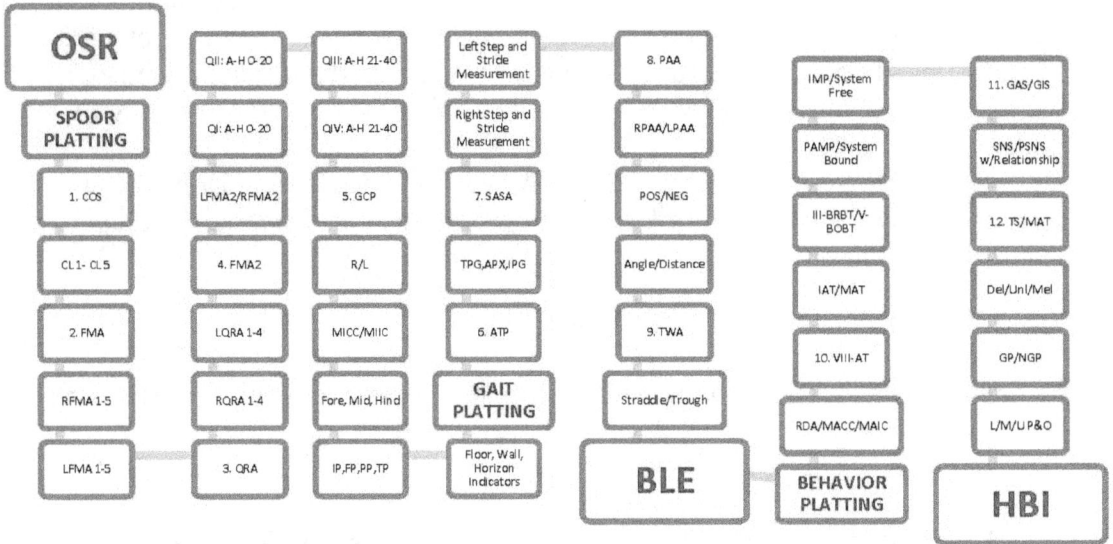

Figure 5.   OCISE Algorithm of Human Behavior
Source: Created by author.

OCISE Algorithm

The OCISE algorithm (see figure 5) has three main sections broken down by OCISE functions. The first function relates to spoor platting (last known spoor/trackers triangle, ground contact points, and aerial travel points), the next function relates to gait platting (gait analysis, gait enablers, and gait deviation), and the final function relates to behavior platting (behavioral analysis). Within the OCISE, functions there are twelve functioning sub-parts defining spoor characteristics, which isolate unique data to individual specificity as well as reliable accuracy as to human behavior identification registered within the spoor-chain signature. They are classification of spoor (COS), foot

58

measurement analysis (FMA), quadrant reference analysis (QRA), foot mapping analysis, micro class characteristics, micro identifying characteristics, spoor dynamics/foot roll analysis, apex of foot arc, stride and step analysis, pitch angle analysis, trail width analysis, macro class characteristics, macro identifying characteristics, trackmaker class characteristics, and innate/manifest adaptive traits.

With the OCISE algorithm being the research instrument used to develop this thesis, this researcher will apply it to five direct visual observation spoor analysis lanes formatted in table 1 to collect the datasets from each. First, to establish precise gait-footfall sequence for baseline establishment; and second, to link the precise gait-footfall sequences collected from each incident reconstruction area to recognizable human behavior identification collected from non-visual observation incident reconstruction area (see table 2). To collect and log the data, a Uniform Scale Format Synch Matrix was developed to record all direct visual observation/non-visual observation gait-footfall sequences of the spoor evidence collected. The Synch Matrix (see Chapter 4-Analysis) contains each function and sub-function of the OCISE algorithm.

The application of the OCISE Algorithm and the logging of the data in the Synch Matrix from both the direct visual observation in the spoor analysis lanes and non-visual observation in the incident reconstruction area will inherently resonate through the component parts within the algorithm, while the gait-footfall sequence and spoor-chain signature continues through both and remains observable to the skill level of this researcher. Namely, these three distinct areas within the algorithm will follow the best practices of observation and spoor recognition methodologies, spoor classification methodologies, and spoor interpretation methodologies.

Table 1.   Spoor Analysis Lane

LANE #	QUARRY	TRAIL TYPE	SLA	GST	GCT	VTP	TEC	DOC %
1	A	A	L	1d	1Sp	1a	Peak 1-	0-10
2	B	A	L	1d	1Sp	1a	Peak 1-	0-10
3	C	A	L	1d	1Sp	1a	Peak 1-	0-10
4	D	A	L	1d	1Sp	1a	Peak 1-	0-10
5	E	A	L	1d	1Sp	1a	Peak 1-	0-10

Source: Created by author.

Table 2.   Incident Reconstruction Area

LANE #	QUARRY	TRAIL TYPE	SLA	GST	GCT	VTP	TEC	DOC %
1	A	I	S	1d	1 Sp	1a	Peak 1-/+	20
2	A	I	S	1d	1 Sp	1a	Peak 1-/+	20
3	A	I	S	1d	1 Sp	1a	Peak 1-/+	20
4	A	I	S	1d	1 Sp	1a	Peak 1-/+	20
5	A	I	S	1d	1 Sp	1a	Peak 1-/+	20

Source: Created by author.

Spoor Observation Methodology

The OCISE process cannot begin without a start point (Last Known Track-LKT/Last Known Spoor-LKS). This means that observation and knowledge skills pre-OCISE are required in order to apply the OCISE Algorithm to any spoor-chain signature. This section describes visual search pattern required to find and maintain a spoor-chain signature following a quarry, but assumes that an initial commencement point has already been established.

Visual Search Pattern

This is a systematic method of searching for a start point in any environment.[2] When using this searching method, the spoor clues reach out to the eyes. Preconceived images upon the environment will be avoided allowing the eyes to observe, classify, and interpret spoor instead of inventing spoor. The method of visual search pattern for this thesis is as follows:

1. Concentrate focus on the farthermost point at, which it is reasonable to identify the quarry. By scanning the farthermost points on each trail using peripheral vision will instinctively alert to observable movement, familiar templates, or colors that are unusual, which are within the line of sight trajectory.

2. Scan the distant areas in the collateral spoor area, paying particular attention to negative spaces and clues that would show the quarry's presence. Imagine all the possible angles, positions and colors the quarry may conform too. Certain attractions will stop the scanning activity and will become compelling drawing visual attention to identify what these interests are, if it is not immediately identifiable, continue to scan the immediate vicinity of the attraction. Alternate from scanning to concentrating on the attraction and project why the attraction is important.

3. Once these farthermost points are searched, concentration on evidence that is closer, working into the extended spoor area, primary spoor area, and if necessary the secondary spoor areas.[3]

By utilizing this method of observation systematic awareness of the surroundings along the trail while varying vision will avoid focus lock. If necessary, objects of interest will be looked at numerous times and will transverse the areas of the tracker's triangle to confirm the signs and spoor along the trail.

Spoor Classification Methodology

Observation within OCISE is constant and extends from the observational awareness, which is pre-OCISE ranging throughout OCISE and finishes after the interpretation prong is complete and OCISE understanding is gained. Classification begins with the last known track.

Spoor Platting

Classification of Spoor

COS has five classes of prints. Class one (CL 1) prints are complete prints with all the details contained within the registration and are very easy to observe. Class two (CL 2) prints are complete prints with little or no details contained within the registration and are easy to observe. Class three (CL 3) prints are partial prints with enough details contained within the registration and are moderate to observe. Class four (CL 4) prints are a portion of a print with no details and only can be recognized as human. Observation of this class of spoor is hard. Class five (CL 5) prints cannot be distinguished as human as they are observed as a sign only. Observation of this class of spoor is severe.[4]

Foot Measurement Analysis

FMA is using a measuring instrument to collect lengths and widths of both right FMA and left FMA CL 1 or 2 impressions. There are five general measurements taken of each foot. The overall length, length of the sole, length of the heel, width of the sole, and width of the heel.[5]

Foot Impression Reference Matrix

Take the foot measurement analysis and apply it to the Foot Impression Reference Matrix (FIRM) (see table 3). This gives a ballpark figure of the height, weight, and shoe size of the quarry followed. The shoe range identifies the range of show sizes to height and weight, height range identifies the height to show size and weight, and the weight range identifies the weight to shoe size and height.

Quadrant Reference Analysis

QRA is the Micro Class Characteristics establishment of four quadrants of both right QRA and left QRA class 1 impressions in order to log the original manufacturing class by confirmation along the spoor-chain signature thus identifying match continuity from one spoor-chain signature for deconfliction purposes from all others. This gives eight total quadrants for continues confirmation. The tracker should be able to take the original drawing of the class 1 prints and while following the spoor-chain signature match all future class 1-3 prints with the original thus piecing the quadrants together like a puzzle for exact spoor-chain signature confirmation.

Table 3.   Foot Impression Reference Matrix

Height in Inches	MEN (70%)			WOMEN (70%)		
	Ages 20-60	Shoe Size Range	Shoe Print Impression	Ages 20-60	Shoe Size Range	Shoe Size Impression
58	N/A	N/A	N/A	102-125	N/A	N/A
59	N/A	N/A	N/A	105-127	N/A	N/A
60	122-136	5 ½-7	9 ½-9 ¾	108-130	4-6 ½	8 ¾-9 ¼
61	125-139	6-7 1/2	9 ¾-10 1/8	112-133	4 ½-7	9-9 ½
62	128-142	6 ½-8	9 ¾-10 3/8	115-136	5 ½ -7 ½	9 ¼-9 ¾
63	132-145	7-8	9 ¾-10 5/8	118-140	6-7 ¾	9 ¼-10
64	136-149	7-8 1/2	9 ¾-10 7/8	121-141	7-8	9 ½-10 1/8
65	139-153	7-9	9 ¾-11 1/8	125-148	7-8 ½	9 ¾-10 3/8
66	142-157	7-9 1/2	10-11 ½	129-152	7 ½-9	9 ¾-10 7/8
67	145-162	7 ½-10	10 ¼-11 ¾	132-156	7 ½-9 ½	10-11 ¼
68	149-166	7 1/2-10 1/2	10 5/8-12	136-160	8-10	10-11 ½
69	153-170	8-11	10 7/8-12 3/8	140-164	8 ½-10 ½	10 ¼-11 ¾
70	157-175	8-11 1/2	11 1/8-12 5/8	144-169	9-11	10 5/8-12
71	161-180	8-12	11 ½-12 7/8	149-174	9-11 ½	11-12 ¼
72	166-185	9-12	11 ¾-13 1/4	154-180	10-12	11 ½-12 5/8
73	170-189	9-12 1/2	12-13 ½	N/A		
74	174-194	9 ½-13	12 3/8-	N/A		
75	178-199	10-13	12 5/8-	N/A		
76	181-205	10-14	12 7/8-	N/A		
77	184-209	11-14	13 ¼-			
78	188-213	12-14 1/2	13 ½-			
79	193-217	12 ½-15				
80	199-221	13-15				

Source: Author created chart based on data obtained from District of Alaska, U.S. Marshals Tactical Tracking Unit (TTU) operational manual.

Foot Mapping Analysis

Foot Mapping Analysis (FMA2) is the Micro Identifying Characteristics, which are randomly branded into footwear through individual use, which damage alters the size, shape, orientation, and position of the original manufacturing. Note type of tread and anything unusual, which will give individual characteristics to the track such as; patterns, which indicate wear on the bottom of the shoes constant tread disruptions (cuts, abrasions, something lodged etc.) edges sharp or worn? (may indicate new or old footwear) does subject drag feet?, walk with his toes, or pound his heels?

Ground Contact Points

Tom Brown Jr. expresses the Ground Contact Points (GCP) this way. "Pressure releases are the visible deformities within and around a track left either by the pressure of the . . . foot as it was making the track, or by the release of that pressure as it lifted."[6] The GCP has two categories indicator pressure release (IPR) and fluctuating pressure release (FPR).[7]

IPR–This GCP category indicates quarry identity and/or psychomotor behavioral programming condition. For instance, finding a class 1 print would be an IPR that the spoor was human. Further detail as to the condition of the human spoor would indicate that the human was male or female, that they had a leg injury, that they are left or right eye dominant, or perhaps that the human has a phobia (fear), is out of control, is very calculating, etc.

FPR–This GCP category indicates quarry locomotor behavioral programming action. For instance, establishing the baseline through spoor platting; to include, all variable trail patterns walking, running, stopping, turning, falling, milling, hesitancies, acceleration, deceleration, etc.

Track Wall-tells direction

Cliff   occurs at the sides of the foot, indicates travel in a straight line.

Ridge   caused by pressure exerted to one side of the track, as in a turn.

Plate   high pressure against the wall definite change of direction.

<u>Wall Explosion</u> sharp turn with highest degree of energy.

<u>Cave</u> usually in front of track, sudden stop or jump to one side.[8]

Track Floor–shows acceleration

<u>Wave</u> small arch in the floor of the track created by the heel striking the ground, compressing soil forward, the foot rolls forward and pushes off with the toe again compressing soil to the center arch of the track. Indicates a stroll!

<u>Double Wave</u> same as a disk but two arches may form. Indicates a dedicated walk.

<u>Disk</u> when walking the toes will tend to pile up loose soil within the front portion of the track, as the subject accelerates, the pressure transfers to the ball of the foot, which in turn transfers any loose soil farther back in the track. The farther back this disk is from the front of the track the faster the individual is accelerating. Indicates fast walk or jog.

<u>Dish</u> once a toe disk passes the center of the track, it is referred to as a heel disk. Indicates fast jog or run.

<u>Explosion</u> caused by so much force that loose soil is cast out of the track, creating a huge plume. Indicates a spring or jump.

<u>Inter-print Pitch</u> overall lengthwise angle of the track: Even, forward and backward pitch Shows body posture: slumped, erect and alert, etc.

<u>Roll</u> overall lateral angle of the track: no roll, right & left roll. A roll-will indicate a turn in the direction of the pitch.

<u>Plume</u> dirt spread out beyond the track: in front fast gait in direction of plume, behind indicates rapid acceleration, circular pattern indicates sudden pivot.[9]

The GCP analysis is invaluable to spoor platting. It will indicate a change of direction before it is made, show a change in acceleration and help speed up the tracking process. The only drawback is an excessive interest can slow down the tracking process.

All knowledge is in the last known track or last known spoor, when it is read correctly, the next step and meaning of the spoor-chain signature is observable and classifiable. In figure 5, the track slope topography illustrates the interplay between the spoor created by the foot through human locomotion and path the foot takes after it leaves the ground traveling in the air between steps. The critical points contained in the aerial travel points is the terminal point when the foot leaves the ground traveling upward to an apex and then traveling downward to an impact point where the foot again contacts the ground and accepts weight sufficient to support the rest of the body during the gait cycle.

Paul Vallandigham presented a paper at the 1999 International Society of Professional Trackers (ISPT) symposium, which brought to light his research of pitch angle and stride length as related to eye dominance and right and left handedness.[10] "Right eye dominant (RED) people will take a longer stride with their right foot than their left–about one inch when walking normally. For left eye dominants (LED), the situation is reversed. Likewise, RED's will point the line of travel with their right foot, creating a narrower pitch angle with their right foot than with their left. The opposite is true of LED's. Since there is such a strong correlation between eye dominance and right or left handedness, this fact is useful in excluding suspects."[11]

Stride and Step Analysis

The spoor-chain signature, which is the signature of the gait cycle scripted into the substrate, contains two measurements. The step is the distance between each foot (opposite feet or left to right or right to left) and is measured from the terminal point of one foot to the impact point of the other foot in succession. The stride is the distance between the same foot (right to right and left to left) and is measured from the terminal point of one foot to the impact point of the same foot in succession.

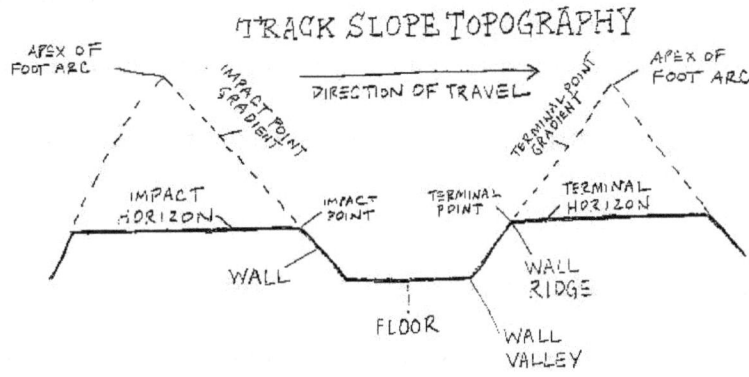

Figure 6.   Track Slope Topography
Source: Created by author.

Pitch Angle Analysis

Each foot as it makes contact with the ground contains an angle distance from the centerline of the body. The pitch of each foot is either a positive number contained in a splaying to the outside of a true center axis or a negative number contained in a pigeoning to the inside of a true center axis. A foot that is place either splayed or pigeoned is considered neutral or on axis. Therefore, the pitch is the angle of foot placement in relation to the spoor-chain signature.

68

Trail Width Analysis

Two widths are contained in the spoor-chain signature. The inner width is called the straddle and is measured from the inner most points of each registered foot from right side to left side. The outer width is called the trough and is measured from the outer most points of each registered foot from right side to left side.

Baseline

The culmination of the observation and spoor recognition should be a very detailed Baseline (BLE) of the spoor-chain signature of the quarry. Without a detailed construction of BLE, it will be difficult to see register deviation anomalies from, which to glean psychomotor behavioral programming. In a detailed government report, *Border Hunter Research Technical Report,* published July 2010, it suggests the importance of establishing the BLE of a quarry in order to understand behavior changes from the proven norms:

> Tracking is about identifying the baseline, perceiving anomalies, and then interpreting these factors to create a cohesive story about what took place. Footprint patterns . . . trash, and other spoor all tell a piece of the story.[12]

> The BLE is pivotal to the overall sign story and requires the tracker, to integrate all the evidence available to predict the quarry's path and intent. Trackers refer to this as "getting into the head of the quarry."[13] A reasonable understanding of the mindset of the quarry will allow the tracker to more accurately interpret the cues available, and to predict the quarry's movements and actions. . . . The tracker must consider the type of quarry. The quarry may be a victim, in, which case the tracker might expect to see blood, vomit, or other signs of a struggle. There may indicators suggesting that the victim is a hostage. The victim may be dead or incapacitated. The victim could be a person who is lost in the wilderness. In other cases, the quarry may not be a victim but rather a criminal or other type of adversary. The tracker will be considering the likelihood that the quarry is armed or unarmed. For an unarmed quarry, the tracker will assess whether the quarry is using mistracking tactics to increase the time-distance gap, avoid detection, and evade capture. He will also

look for signs of panic on the part of the quarry including irrational behaviors and a lengthening stride.[14]

BLE gives the tracker the mind of the quarry. The mind is the action arm of the brain,[15] and therefore by employing the methodology of the "O" and "C" in the OCISE algorithm, the tracker gleans the BLE. Once the BLE is established, then accuracy can be attained in the interpretation of human behavior, which is manifest through body actions.

Spoor Interpretation Methodology

Behavior Platting

Once the tracker has collected the data from the spoor-chain signature, the baseline is imbedded visually within the trackers mind. The baseline indicates the locomotor behavioral programming of the quarry and is used to recognize any psychomotor anomalies. These anomalies are considered register deviation anomalies and represent psychomotor behavioral programming of the quarry. The tracker should use the following simple equation to find the quarry human behavior identification:

$$BLE\ (Input) + BLE\ (output) + RDA = HBI$$

Hunter B. Armstrong, in an article he wrote for the International Hoplology Society titled *Pre-Arranged Movement Patterns* states:[16]

> Patterns of human movement can be roughly divided into two categories, A–those derived from a genetic base, and B–those that must be learned, often called "skills." . . . the A types, "primary movement patterns," and the B types, "secondary movement patterns." Both types of movements can range from very simple actions to complex patterns of movement. Blinking would be an example of a *simple* primary movement pattern, where a very basic reflex action occurs; an example of a *simple* secondary movement pattern is winking with one eye–here the movement is also relatively simple, but must be learned and practiced. Examples of *complex* movements would be walking (primary), which is not learned, but naturally develops as greater motor control arises in the maturing nervous system; and dancing (secondary), which though based on walking,

requires training to learn the combined movements and to enhance motor control.[17]

Descriptions of Loco-Motor Behavioral Programming

Normal Walking Gait (NWG) is considered a primary movement pattern, which is complex due to the maturing of the nervous system during development. Since humans choose innately the normal walking gait for conservation of energy during locomotion, it will be used in this thesis for the methodology base.

Human Loco-Motor Behavioral Programming Patterns

Normal Walking Gait (NWG)–NWG is the primary movement pattern of all humans. The NWG primary movement pattern is established by the presence of a plume; a low angle single, long wave; little or no toe dig; little or no registration at the impact point and terminal point. Ground Contact Point indicator pressure releases usually mark well in the heel but may not toward the front of the foot as the sole of the foot flexes.

Loco-Motor Behavioral Programming Register
Deviation Anomalies to Normal Walking

Fast Walking–Leaves a double wave; deep toe dig; short plume in front; higher angle of entry and exit (because feet are picked up more).

Backward Walking–Quarry avoiding capture may walk backwards in an attempt to fool the tracker. Backing tracks have no toe dig (but they may have a heel dig); the pattern of the sole actually shows in the toe and front of the foot; any plume will be at the heel instead of the toe; and usually the stride is short, straddle is wide, and the line of travel is not straight.

Uphill Walking–Quarry walking up hill will dig the toes into the hill to create a plateau or shelf for stability for the next step. Oft times there are slip marks where the soil gives way and gravity slides the foot down the heel until the soil catches. In addition, the edges of the foot may be employed for stability.

Downhill Walking–Quarry moving downhill has some of the same problems as uphill walking. Going downhill, however, the heel is the primary contact with the soil. The quarry will dig the heel in to establish the plateau. Again, sliding and falling are very common. The side or edge of the foot is used to also give perpendicular support for the descent.

Stagger Walking–This form of movement is usually unintentional and is caused by exhaustion or injury. The gait is erratic and the foot is drop, dragged, and is place on the ground with a floor ground contact points that is pitched. This produces an unstable based that cannot support the body and the stagger movement begins to try to catch the body. Unfortunately, when the quarry gets to this point, hustle is necessary from the tracker.

Running

In running, the front of the track is usually very churned up, but you may see a pattern in the disk

Running (General term for fast moving)–Noticeably longer stride; one gentle wave in back and a taller, overlapping one in front; longer plume; and the front mogul may be exaggerated by the fact that part of it has been moved back as one piece, called a disk. The faster the run, the farther back the disk will be. With quick acceleration there will be a plume going backwards.

Descriptions of Psycho-Motor Behavioral Programming

All outward movement is a stamp of an inward thought or emotion, and therefore registers within the spoor-chain signature all the emotions and thoughts as ground contact point indicator pressure releases or fluctuating pressure releases. Psychologist William James aptly observed,

> Objects of rage, love, fear, etc., not only prompt a man to outward deeds, but provoke characteristic alterations in his attitude and visage, and affect his breathing, circulation, and other organic functions in specific ways. . . . As with instincts, so with emotions, the mere memory or imagination of the object may suffice to liberate the excitement. One may get angrier in thinking over one's insult than at the moment of receiving it; and we melt more over a mother who is dead than we ever did when she was living.[18]

Dr. Dispenza eloquently asserts that: "Your every thought produces a biochemical reaction in the brain. The brain then releases chemical signals that are transmitted to the body, where they act as messengers of the thought."[19]

It matters not whether the outward movements of the quarry are prompted by instincts, emotions, or thoughts; if the human body acts upon the innervations, the biomechanical system must accommodate the impulse. Therefore, in either primary movement patterns or secondary movement patterns the eight adaptive traits are manifest as psychomotor behavioral programs. The ground contact points register the indicator pressure releases and fluctuating pressure releases as an emotion or thought encrypted into the spoor-chain signature waiting for only the tracker to observe, classify, and interpret.

Human Psycho-Motor Behavioral Programming Patterns

As human beings we attached our thoughts to emotions converting them into motivation to act upon the chemical impulses produced from the brain. To understand psychomotor behavioral programming is to understand the connection between the locomotor apparatus of the body as the application endstate of our brain, mind, thought, and emotion. Ronald de Sousa shows how broad the emotions platform is in its effect on the human experience:

> A broad consensus has emerged on what we might call adequacy conditions on any theory of emotion. An acceptable philosophical theory of emotions should be able to account at least for the following nine characteristics.
>
> - Emotions are typically conscious phenomena; yet
> - They typically involve more pervasive bodily manifestations than other conscious states;
> - They vary along a number of dimensions: intensity, valence, type and range of intentional objects, etc.
> - They are reputed to be antagonists of rationality; but also
> - They play an indispensable role in determining the quality of life;
> - They contribute crucially to defining our ends and priorities;
> - They play a crucial role in the regulation of social life;
> - They protect us from an excessively slavish devotion to narrow conceptions of rationality;
> - They have a central place in moral education and the moral life.[20]

Emotions play a pivotal role in the study of register deviation anomalies as well as in the concept of baseline within OCISE. Although, linearity gives structure to the science of spoorology as a quantitative tool in baseline development specifically; the qualitative analysis inherent in non-linearity cannot be glossed over in the interpretation of spoor evidence and human behavior in general. In its strictest sense, although both James and Dispenza make clear that although the body must act out the neural programming of thought within a quarry, each thought is usually attached to an emotion for the body to sense or feel through the neurosynaptic network. Dr. Dispenza further asserts:

> The thoughts that produce the chemicals in the brain allow your body to *feel* exactly the way you were just *thinking*. So every thought produces a chemical that

74

is matched by a feeling in your body. Essentially, when you think happy, inspiring, or positive thoughts, your brain manufactures chemicals that make you feel joyful, inspired, or uplifted . . . when you anticipate an experience that is pleasurable, the brain immediately makes a chemical *neurotransmitter* called *dopamine*, which turns the brain and body on in anticipation of that experience and causes you to begin to feel excited. If you have hateful, angry, or self-deprecating thoughts, the brain also produces chemicals called *neuropeptides* that the body responds to in a comparable way. You feel hateful, angry, or unworthy.[21]

Taking the holistic approach proffered by Huitt in his systems model of human behavior and applying it to spoorology uncovers the secret to interpretation through behavioral platting within OCISE. The mind is composed of cognition (knowing, understanding, thinking), affection (feelings, emotions, attitudes, predispositions), and conation (volition, will, intentions to act, reasons for taking action), which collects information and manifests action through the physical body. The physical body is genetic influences, physical functioning, and what he calls "overt behavior." Also, contained in this holistic approach is a feedback loop between overt behavior and the resulting stimuli from the environment.[22] The mind of any quarry, as action (behavior) is implemented, is scripted to the medium of the soil and vegetation by gravity impressions of the feet during locomotion.[23] The key to both locomotor behavioral programming and psychomotor behavioral programming is the direct connection between output and input as observed within all three components of OCISE.

The following are two examples of observable psychomotor behavioral programming that stem from the mind of a quarry to the spoor-chain signature:

Grief/Weariness/Fatigue/Labored Cognition

The movements are made slowly, heavily, without strength, unwillingly, and with exertion, and are limited to the fewest possible. Outwardly then, the quarry walks slowly, unsteadily, dragging the feet and hanging the arms. The knees are unsteady in support of locomotion, and with this condition of weakness of the voluntary thoughts of the neocortex comes the degradation of energy to move the muscle apparatus of the whole body. This condition brings with it a motor weakness, which drags the feet along the ground with atypical shortened gait. Accompanying, the weakness are frequent stops with no apparent reasoning except to suggest a mindset, which sluggishly innervates the psychomotor apparatus through the midbrain acting on established neural patterns that the frontal lobe cannot override. Voluntary movement is unwanted and the quarry only moves to fight the cause of the weariness. Rather than expend energy to walk, the quarry would rather sit in a sunken state looking inward instead of outward at the world around and reflexes are slow when responding to external stimuli through the senses.

Panic/Fear (Phobia)/Out of Control/Without Cognition

The frightened quarry, at first, stands motionless holding breath to avoid the object of fear. Perhaps, the quarry will crouch down cowering away from the object hoping that the object of fear does not recognize the quarry's presence. A cold sweat may appear do to clammy pale skin. The feet revolt and follow the fear with short sharp cuts, which move away from the object. As the fear escalates to terror, the resultant explosion of violent emotions sparks many possible overt behaviors. All the muscular apparatus may become rigid on the one hand or explode into convulsive scurry of movement, which scampers the feet erratically away from the object of fear. The baseline is suddenly aborted in confusion as the body tries to recover from the object. Shortened steps and stride, wider trail width, and the ground contact points are labored by foot drags to maintain stability. Eventually, escape is manifest as the spoor-chain signature begins to open up through locomotor behavioral programming, trying to reestablish a baseline, which is conducive to speed. Gradually from the confusion of clustered movement comes an increased distance within steps and stride, the trail width begins to narrow, and the ground contact points recover from the extreme angles and damage to the wall, floor, and horizon.

Eight Adaptive Traits of LMBP and PMBP

The adaptive traits assist innately both locomotor behavioral programming and psychomotor behavioral programming. The interaction of the adaptive traits within the context of human locomotion substantiates the totality necessary for the applied synergy to manifest adaptive traits during locomotion. As the BGS paradigm lays the science for spoorology contained in linearity and laws of motion. The OCISE algorithm structures and frames in order to adhere to a holistic approach for the application of science to spoor evidence. The eight adaptive traits network but do not intersect. Thus, the interface of the eight adaptive traits become greater than the totality of all quantities. Eight adaptive traits manifestations; therefore, exist physically in the primary movement and secondary movement patterns.[24]

Manifest Adaptive Traits

The manifest adaptive traits refer to the behavioral expression of malleable traits of existence, which develop throughout the bio-social evolution of the quarry. The manifest adaptive traits, as expressions of neuro-physiological-anatomical processes, are locomotor and psychomotor presentations of the probable intrinsics of and in each quarry. The manifest adaptive traits are mediated either by primary movement or secondary movement patterns (system bound or system free).[25]

Three Brain-Bound Traits

Overview

All environmental perceptions are received by the mind through the casting of the senses. Since the environment is not stagnant, thus ever changing, the natural interaction of the physical body to the environment is monitored through the mind by the brain. The mind's conation (MVT) is reconciled by cognition (MCIT). It is the study of this process in determining quarry action tonification or action sedation given interaction with the environment, which implicates overt behavior congruent with the mind's intent as it is engraved in the substrate.

In unification of cognition and conation, intent is made possible through affection (MIST). As the senses cast and the mind perceives environmental factors, the alarm reaction functions to initiate the general adaptive syndrome should it be necessary. The general adaptive syndrome prepares physically for survivability through "fight or flight" response. As alarm reaction is interdicted or inhibited by the override of the neocortex through special hormones to block precipitation, when completely blocked imperturbable-mind state is achieved; when partially blocked steadfast-mind state is achieved. When the complete failure to block alarm reaction is manifest the mind produces fear, panic, and rage. Achieving imperturbable-mind state or steadfast-mind state, the mind is adaptive. Without imperturbable-mind state or steadfast-mind state, the mind is maladaptive.[26]

Manifest Cognitive/Intuitive Trait

The manifest cognitive/intuitive trait suggests that the brain perceptively collects data from the environment for, which the quarry is imbedded and evaluates what is collected by the complex interaction between the left and right hemispheres of the brain. The manifest cognitive/intuitive trait perceives the environment and negotiates and resolves complexities of human movement both cognitively or intuitively.[27]

Manifest Volitional Trait

The manifest volitional trait is learned locomotor and psychomotor behavioral programming based on the life-experiences and/or conditioned training of the quarry. manifest volitional trait concepts are "determined by genetic/constitutional/ temperamental factors, and . . . presided over by"[28] the manifest imperturbable mind/steadfast-mind trait. Manifest volitional trait is the nexus between sensory-neural functions and the gait-footfall sequences effect in the environment.[29]

Manifest Imperturbable Mind/Steadfast-Mind Trait

Manifest imperturbable mind/steadfast mind trait is the glue, which binds manifest cognitive/intuitive trait and manifest volitional trait and solidifies the effective use of both within the context of human locomotion. In manifest imperturbable mind/steadfast mind trait, both innate and manifest are imbedded deep within human inherent instinct for survival. This perpetuates the manifest imperturbable mind/steadfast mind trait induction through the neo-cortex for interdiction of the alarm reaction during environmental perceptions, which invokes either imperturbable mind state or steadfast mind state to maintain the baseline through the spoor-chain signature or to deviate based on manifest volitional trait.[30]

Two Body-Bound Traits and Three Action-Bound Traits

<u>Overview</u>

Both body-bound traits and action-bound traits are manifest through the primary movement patterns and secondary movement patterns of locomotor and psychomotor behavioral programming. They achieve optimal integration and therefore concentration for all loco and psychomotor behavioral programming through interfacing abdominal tensing with abdominal breathing.

Manifest Abdominal Trait

Manifest abdominal trait unites the torso consolidating the bio-mechanical exertions of the locomotor platform and the head, arms, and torso.[31]

Manifest Respiratory/Vocality Trait

Manifest respiratory/vocality trait assists with optimizing bio-mechanical exertion with manifest abdominal trait thus achieving unity of effort in moving the locomotor platform and the head, arms, and torso.[32]

Manifest Omni-poise Trait

Manifest omni-poise trait is the point of origin for all human locomotion. It is exemplified through the natural posture of standing upright. Both stability and mobility intersect at the natural posture and forms the baseline for all locomotion. Stability or standing in place is seen as potential energy from, which kinetic energy will emerge. Mobility conversely characterizes the change from potential to kinetic energy with, which human movement progression is stimulated. Mobility also suggests a return to stability should either kinetic energy deplete or motive suggest conservation due to mobility interaction with gravitational forces for, which stability mediates balance when displacement alters locomotor or psychomotor behavioral programming.[33]

Manifest Force-Yield Trait

This trait interacts with manifest omni-poise trait by investing in primary and secondary movement patterns suggesting an interplay between applying force or yielding to force in conjunction to quarry objectives (strategic, operational, tactical).[34]

Manifest Synchronicity Trait

This trait is a design manifestation of primary and secondary movement patterns suggesting interaction of time and space congruent to achieving an objective. There is a coordination sequencing when manifest omni-poise trait, Manifest force-yield trait are implemented with manifest synchronicity trait. In order to synchronize bio-mechanical expressions an understanding of locomotor platform and head, arms, and torso structures and functions in relation to the quarry objectives must be anticipated.[35]

General Adaptive Syndrome/General Inhibition Syndrome

The autonomic nervous system consists of both the sympathetic nervous system and the parasympathetic nervous system represented by GAS and GIS.

General Adaptive Syndrome

The "fight or flight" response to object stress was first described by Hans Selye (figure 6) when he built a three phase process from the initiation phase of the response in the AR to the final stage of energy depletion, exhaustion, and collapse.[36] The three phases are:[37]

> In the first stage of GAS called the alarm reaction, the body releases adrenaline and a variety of other psychological mechanisms to combat the stress and to stay in control. . . . The muscles tense, the heart beats faster, the breathing and precipitation increases, the eyes dilate, the stomach may clench . . . this is done by nature to protect . . . once the cause of the stress is removed, the body will go back to normal.

> If the cause of the stress is not removed, GAS goes to its second stage called resistance and adaptation. This is the body's response to long term protection. It secretes further hormones that increase blood sugar levels to sustain energy and raise blood pressure. The adrenal cortex (outer covering) produces hormones called corticosteroids for this resistance reaction. . . . If this adaptation phase continues for a prolonged period of time without periods of relaxation and rest to counterbalance the stress response, sufferers become prone to fatigue, concentration lapses, irritability and lethargy as the effort to sustain arousal slides into negative stress.

> The third stage of GAS is called exhaustion. In this stage, the body has run out of its reserve of body energy and immunity. Mental, physical and emotional resources suffer heavily. The body experiences "adrenal exhaustion." The blood sugar levels decrease as the adrenals become depleted, leading to decreased stress tolerance, progressive mental and physical exhaustion . . . and collapse.

As described above, in the short term, by sympathetic nervous system innervation and hormonal dumping with the precipitation of the alarm reaction. The augmented parameters are adaptive in the short term but maladaptive in the long term.

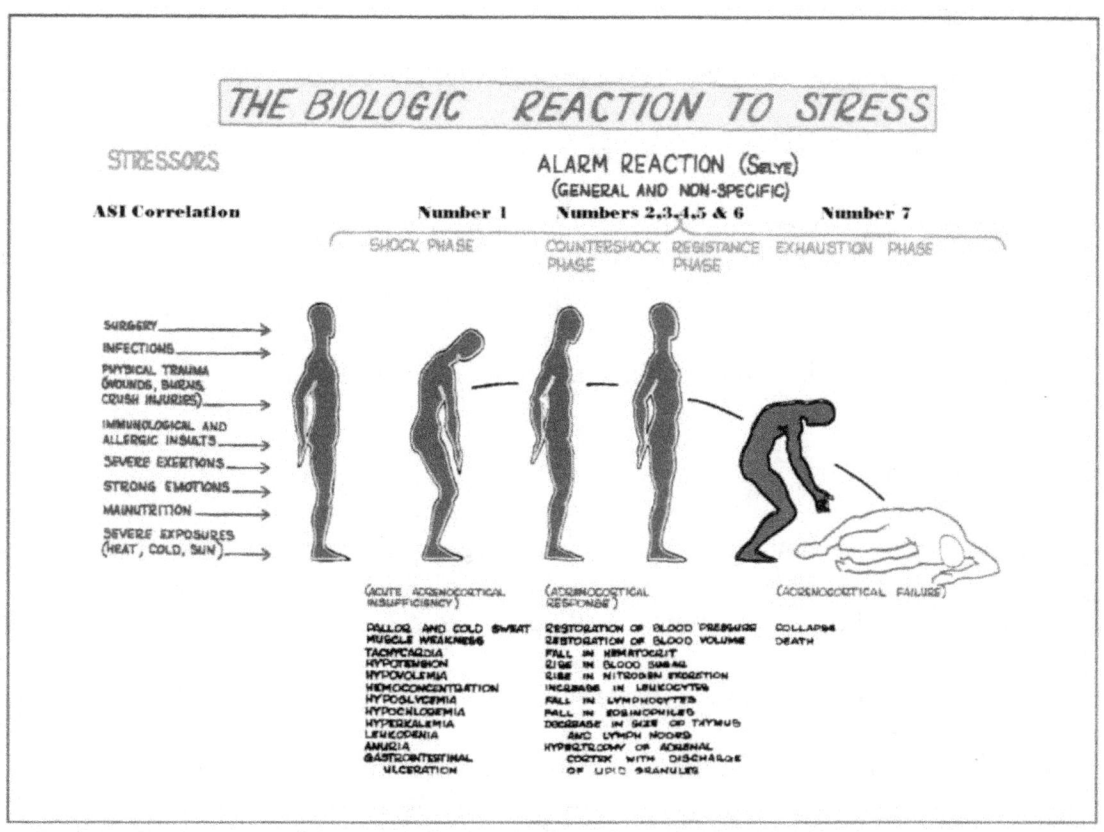

Figure 7.　Alarm Reaction

Source: Clymer Healing Research Center, http://www.chronicfatigue.org/History.html (accessed 6 April 2011).

Richard Hayes describes how the GAS operates:

With the evolution of the neocortex, and the two hemispheres connected by the corpus callosum, there emerged a neocortical override of the Alarm Reaction (AR), via a hormonal blocking of the initiation/precipitation of the AR in the old brain structures. In other words, seamless with the first perception of danger/threat (via the Cognitive/Intuitive Trait), there is a neocortically derived inhibition or interdiction of the precipitation of the AR. This is accomplished by hormonal blocking in the anatomical loci of the old brain structures. . . . There is a complete (100%) interdiction of the AR (imperturbable-mind), or a partial (less than 100%) interdiction (steadfast-mind), in, which there is hormonal dumping within adaptive limits (i.e., without panic, mindless rage, or mindless fear).[38]

83

General Inhibition Syndrome

In research conducted by the *International Hoplology Society* led by Richard Hayes, they found data to suggest that there is another syndrome, which co-exists based on proportionality. This syndrome is the GIS. The GIS is operated by the parasympathetic nervous system, which function is to decrease the physiological and metabolic parameters, which produces adaptation within alarm reaction.

GAS/GIS Co-Existence

The axis between both the GAS and GIS suggests the normal physiological and metabolic parameters both during repose and during exertion. Since the body must maintain a homeostatic environment for survival existence within the locomotor and psychomotor behavioral programming, it is not surprising that there is both input and output, which exists between the mind and body; between the body and environment; and, between mind and environment. In both GAS and GIS, the quarry can either be adaptive or maladaptive should the traits fail to function to standard. This is projected through the locomotor apparatus of gait-footfall sequences and registered into the substrate by the spoor-chai signature.

Transcendent Synergy of the Manifest Adaptive Traits

All psychomotor behavior are rooted in some intersect of time and place on the substrate of earth through the gait-footfall sequences of the spoor-chain signature. The psychomotor behavioral programming registered in any spoor-chain signature is a course-of-action of the trackmaker based on the science of the BGS paradigm through linearity and laws of motion. Through OCISE, the tracker is able to see and understand the historical record left encrypted in a crime scene. The manifest adaptive traits are universally constant within all humans. The practicality of manifest adaptive traits, regarding the experience of a trackmaker and the behavioral aspect of performance allows the tracker to analyze the component traits manifest during locomotor and psychomotor behavioral programming within the spoor-chain signature, and to evaluate their implications, both individually and holistically to a crime scene. Each tracker can appraise and place value on their own performance, thereby reading the performance of others in the spoor-chain signature context to the psychomotor behavior. A tracker does this experientially, from an introspective (output), looking inward from the spoor-chain signature back to the mind of the trackmaker, while the tracker in the midst of analyzing a spoor-chain signature, or behavior, looking reflectively (input) outward, observing the performance of the trackmaker laying the spoor-chain signature from the mind of the tracker as the baseline is solidified.

Human Behavior Identification

Human behavior identification is achieving the sign story as to either locomotor or psychomotor behavior through OCISE algorithm application. All primary and secondary movement patterns are contained within both locomotor and psychomotor behavior. Utilizing the eight adaptive traits through the manifest adaptive traits, once a baseline is attained, allows the tracker to understand the mind of the quarry through the gait-footfall sequences registered in the spoor-chain signature. This leads the tracker through experience of scientific, technical, and specialized skills to the intent or "mind" of the movement to gain proportionality, additivity, output to input constancy, with conclusive and deterministic extrapolations. Any primary and secondary movement

85

pattern indicator pressure releases or fluctuating pressure releases is unique in its pattern to a trackmaker and establishes the gait-footfall sequences and its subsidiaries of movement implicating the mind to the spoor-chain signature.

Strengths of OCISE

The strength of the OCISE method is that it provides a linearity approach to understanding the spoor data collected from a crime scene. Mathematically measuring spoor data allows this researcher to draw conclusions, which are more accurate and reliable when answering the primary and secondary research questions. Additional strengths of this approach is the universal applicability of OCISE to any crime scene, in any jurisdiction, anywhere in the world based on the new BGS paradigm.

---

[1]Albert "Ab: Taylor and Donald C. Cooper, *Fundamentals of Mantracking: The Step-By-Step Method,* 2nd ed. (Olympia, WA: ERI International, 1990), 37.

[2]Bob Carss, *The SAS Guide to Tracking* (New York: Lyons Press, 2000), 66.

[3]Ibid.

[4]Taylor and Cooper, 55.

[5]Robert Speiden, *Foundations for Awareness, Signcutting and Tracking* (Christiansburg, VA: Natural Awareness Tracking School, 2009), 40-41.

[6]Tom Brown, *Tom Brown's Field Guide to Nature Observation and Tracking* (New York: Berkley Books, 1983), 209.

[7]Ibid., 209-210.

[8]Ibid., 211-221.

[9]Ibid.

[10]Paul Vallandigham, "Eye Dominance and Your Body" (Paper presented at the Annual Symposium of the International Society of Professional Trackers, Petaluma, CA, 22-24 October 1999).

[11]Paul Vallandigham, "Tracking: 19th Century Art as 21st Century Science" (Paper presented at the Annual Symposium of the International Society of Professional Trackers, Petaluma, CA, 22-24 October 1999).

[12]David Fautua et al., *Border Hunter Research Technical Report* (Norfolk, VA: U.S. Joint Forces Command, 2010), 11.

[13]Ibid., 108.

[14]Ibid., 109.

[15]Joe Dispenza, *Evolve Your Brain: The Science of Changing Your Mind* (Deerfield Beach, FL: Health Communications), 61.

[16]Hunter B. Armstrong, "Pre-Arranged Movement Patterns," *HOPLOS* 6 (Winter 1988): 18-21.

[17]Ibid.

[18]William James, *The Principles of Psychology: Volume Two* (Mineola, NY: Dover Publications, 1950), 442.

[19]Dispenza, 43.

[20]Ronald de Sousa, "Emotion," *The Stanford Encyclopedia of Philosophy* (Spring 2010), http://plato.stanford.edu/archives/spr2010/entries/emotion/ (accessed 6 April 2011).

[21]Dispenza, 43.

[22]Huitt, 2.

[23]W. Huitt. "A Systems Approach to the Study of Human Behavior," *Educational Psychology Interactive*, http://www.edpsycinteractive.org/materials/sysmdlo.html (accessed 16 September 2010).

[24]Richard Hayes, *Paleolithic Adaptive Traits and the Fighting Man* (Sedona, AZ: International Hoplology Society, 1998). The EAT used by hoplologists of the IHS is the base for the underlying behavior in spoor evidence. Where acronyms have been established by IHS for these terms the author uses them. Where no acronyms are established already, the author has created them.

[25]Ibid., 4.

[26]Ibid., 12.

[27]Ibid., 6.

[28]Ibid.

[29]Ibid., 7.

[30]Ibid., 5.

[31]Ibid., 8.

[32]Ibid., 8-9.

[33]Ibid., 7.

[34]Ibid., 9.

[35]Ibid., 10-11.

[36]Ibid., 19.

[37]Holistic Online.com, "Stress: The Silent Killer," http://holisticonline.com/stress/stress_gas.htm (accessed 6 April 2011).

[38]Ibid., 20.

CHAPTER 4

ANALYSIS

Tracking evidence relates to a uniquely developed visual examination and recording of evidence that is most often otherwise overlooked. Trackers are specially trained to visually locate, collect, preserve and present latent footprint evidence that is not obvious, or clearly visible to experienced crime scene investigators. Inherent with tracking examinations is the training, experience and skill to accurately interpret the movement of persons, their physical and mental state, and individual characteristics from the footprint evidence.[1]

— Joel Hardin with Matt Condon, *Tracker: Case Files and Adventures of a Professional Mantracker*

Overview

The last chapter presented the OCISE algorithm as criteria methodology, based on the science of linearity and laws of motion within the BGS paradigm, to assess the extent to, which spoor evidence will elucidate human behavior thus addressing the primary and secondary research questions. This chapter will breakdown the collected data from the OCISE algorithm to answer the primary and secondary research questions as to whether human behavior can be gleaned from the spoorological record.

Spoor Analysis Lane

Purpose

To glean sufficient spoor evidence from each trackmaker through direct visual observation to form the baseline through OCISE.

Construction

Spoor Analysis Lane (SAL) was constructed within an outdoor horse-riding arena on Fort Leavenworth. The full width of SAL was twenty (20) meters from the outer most points left to right. The length of each lane was approximately twenty-five (25) meters from start to finish.

Conditions

The temperature was 45 degrees Fahrenheit, no precipitation, with a stiff wind making it necessary for all trackmakers to wear layers. The slope topography analysis was flat. Ground hardness type was soft fine soil covering a hardened layer of packed clay.

Design

Each lane was four (4) meters wide and the trackmaker H1 through H5 were assigned lanes with a distinct start point and ending point marked visually by this researcher. Each trackmaker walked down the center of their respective lanes so that there was approximately four (4) meters between them.

Evidence Recovery Projections (Physical Evidence)

SAL would produce:

1. One (1) gait-footfall sequence locomotor behavioral program from each lane for a total of five (5).

2. One (1) spoor-chain signature psychomotor behavioral program from each lane for a total of five (5).

3. Thirty (30) to thirty-five (35) separate pieces of spoor evidence from each lane for an approximate total of between one hundred fifty-five (155) to one hundred seventy-five (175).

Incident Reconstruction Area

Purpose

To glean sufficient spoor evidence from each trackmaker through non-visual observation to establish identity of the trackmaker, to re-establish baseline to the spoor-chain signature, and to form human behavior identification through OCISE.

Construction

Incident Reconstruction Area (IRA) was constructed within an outdoor horse-riding arena on Fort Leavenworth. The full width of IRA was twenty-five (25) meters from the outer most points left to right. The length was approximately seventy-five (75) meters from front to back. There were no distinct lanes.

Conditions

The temperature was 45 degrees Fahrenheit, no precipitation, with a stiff wind making it necessary for all trackmakers to wear layers. The slope topography analysis was flat. Ground hardness type was soft fine soil covering a hardened layer of packed clay.

Design

Within IRA, each trackmaker choose the method of gait-footfall sequence to negotiate situations to illicit locomotor and psychomotor behavior naturally. Some situations were done alone and some were done as groups, but they were all non-visual observation to the researcher. The trackmakers were constrained to IRA to conduct each situation and each trackmaker had to participate in the individual situations but they could choose who participated in the group situations. The situations were as follows:

91

1. You were just notified that there has been a death in your extended family of someone that you were very close with. They live out of state and you are walking home.

2. You are taking a hike in an area of the national forest. You lost track of time and it is getting dark. You begin back to the trailhead but are disoriented.

3. You and your family have a marijuana grow operation at a remote location on your private 500 acre property nestled up to a state park. You believe that state law enforcement may be conducting surveillance of your operation, but you are not sure. One day you decide to take the high ground to see if you can spot them following you. You are alone or with one of your family members.

Evidence Recovery Projections (Physical Evidence)

IRA would produce:

1. One (1) gait-footfall sequence locomotor behavioral program from each trackmaker.

2. One (1) spoor-chain signature psychomotor behavioral program from each trackmaker.

3. "Many" separate pieces of spoor evidence from each trackmaker within each situation.

4. Baseline confirmation for identity of trackmaker spoor-chain signature in the situations and their respective sign story.

5. Human behavior identification proportionate to the "mind of the trackmaker."

Classification of Spoor

All prints left in the spoor analysis lane and incident reconstruction area were classification of spoor A1 suggesting that the variable of change in ground cover (GST and GVT) were controlled in order to mitigate the terrain variances. The A1 prints were established for all five spoor-chain signatures within both the spoor analysis lanes and within the incident reconstruction area. From Z1 through Z5 all showed complete detail as to trackmaker identity. The prints were all different because of each trackmaker wearing different manufacturer tread patterns, which distinguished each from the other by applying visual acuity to allow track discrimination. A change in distance between the markers throughout the duration of the experiment allowed for a decrease in unforeseen error due to marker visibility to each trackmaker. Table 4 establishes the consistency of track visibility in both spoor analysis lanes and the incident reconstruction area for all trackmakers Z1 through Z5.

Table 4.   Classification of Spoor

OBSERVATION & SPOOR RECOGNITION (OSR)												
			SAL	IRA	SAL	IRA	SAL	IRA	SAL	IRA	SAL	IRA
LKT/LKS	A-V	OCISE	Z1	Z1	Z2	Z2	Z3	Z3	Z4	Z4	Z5	Z5

SPOOR PLATTING												
1. COS	A											
Class 1 Print	1		X	X	X	X	X	X	X	X	X	X
Class 2 Print	2											
Class 3 Print	3											
Class 4 Print	4											
Class 5 Print	5											

Source: Created by author.

Figures 8 through 12 show visually the differences in the shoe patterns worn by each trackmaker during both the spoor analysis lane and incident reconstruction area respectively. The peculiarities between each trackmakers tread pattern based on index comparative analysis as ground spoor is distinct. Given the parameters within the spoor analysis lane based on uniqueness of all spoor-chain signatures and the barriers constraining normal walking gait to control natural lines drift, the separation alleviated spoor contamination in each lane.

Figure 8.   Z1 Trackmaker

Source: Photo by author.

Figure 9.   Z2 Trackmaker

Source: Photo by author.

Figure 10. Z3 Trackmaker

Source: Photo by author.

Figure 11. Z4 Trackmaker

Source: Photo by author.

Figure 12. Z5 Trackmaker

Source: Photo by author.

Foot Measurement Analysis

Once classification of spoor was established for all trackmakers within spoor analysis lanes and incident reconstruction area. Measurements were taken of each foot placement. Table 5 shows the measurements as broken down by the foot measurement analysis format both for the right foot as well as the left foot. The foot measurement analysis is: overall length, length of the sole, length of the heel, width of the sole, and width of the heel. Notice that Z1 through Z5, all trackmakers do not have the same overall track length with either foot. This distinguishing factor alone confirms the observation and spoor recognition with the spoor-chain signature of each trackmaker. When looking at the data it shows that all other pertinent data is distinguishable from each spoor-chain signature, except for the length of the sole of Z2 and Z3. Also, notice that the heel length of Z3 and Z4 are identical. When taking all data for foot measurement analysis into consideration, foot measurement analysis is quantitatively accurate and segregates all trackmakers within both the spoor analysis lanes and the incident reconstruction area.

Table 5. Foot Measurement Analysis

OBSERVATION & SPOOR RECOGNITION (OSR)													
			SAL	IRA	SAL	IRA	SAL	IRA	SAL	IRA	SAL	IRA
LKT/LKS	A-V	OCISE	Z1	Z1	Z2	Z2	Z3	Z3	Z4	Z4	Z5	Z5

SPOOR PLATTING													
2. FMA	B											
Foot Length	1	R	13.0	13.0	13.5	13.5	12.0	12.0	11.5	11.5	11.75	11.75
Sole Length	2		6.75	6.75	7.0	7.0	7.0	7.0	5.75	5.75	6.0	6.0
Heel Length	3		3.63	3.63	4.5	4.5	4.0	4.0	4.0	4.0	3.5	3.5
Sole Width	4		5.0	5.0	5.0	5.0	4.5	4.5	4.0	4.0	4.13	4.13
Heel Width	5		3.5	3.5	4.25	4.25	3.75	3.75	2.75	2.75	3.25	3.25
Foot Length	1	L	13.0	13.0	13.5	13.5	12.0	12.0	11.5	11.5	11.75	11.75
Sole Length	2		6.75	6.75	7.0	7.0	7.0	7.0	5.75	5.75	6.0	6.0
Heel Length	3		3.63	3.63	4.5	4.5	4.0	4.0	4.0	4.0	3.5	3.5
Sole Width	4		5.0	5.0	5.0	5.0	4.5	4.5	4.0	4.0	4.13	4.13
Heel Width	5		3.5	3.5	4.25	4.25	4.0	4.0	2.75	2.75	3.25	3.25
FIRM	C											
Approx. Height	1	Hgt	72-78		72-78		68-73		68-72		67-72	
Approx. Weight	2	Wgt	166-213		166-213		149-189		136-180		145-185	
Mechanical Step Zero	3	MS Z	17	17	17	17	17	17	17	17	17	17

Source: Created by author.

Quadrant Reference Analysis

The QRA in table 6 allows for micro class characteristics of both right and left QRA for all A1 prints within the spoor analysis lanes and incident reconstruction area. Since confirmation is not contested in the spoor analysis lanes do to both experiment restraints and constraints, a detailed analysis of QRA is not necessary for establishing the baseline or human behavior identification. All spoor analysis lane spoor-chain signatures acquired during direct visual observation were used to match exactly the spoor-chain signatures within the incident reconstruction area through non-visual observation of the spoor-chain signatures. Using OCISE with the deliberate trailing method, all spoor evidence was observed, classified, and recorded. Deliberate trailing method finds every

98

step, excluding no evidence. QRA was utilized to validate identity in the incident reconstruction area. It is important to point out that all QRA must be established within a crime scene in order to deconflict a scene when multiple trackmakers are manipulating the substrate. This briefly happened in situation 3.

Table 6.   Quadrant Reference Analysis

OBSERVATION & SPOOR RECOGNITION (OSR)												
			SAL	IRA	SAL	IRA	SAL	IRA	SAL	IRA	SAL	IRA
LKT/LKS	A-V	OCISE	Z1	Z1	Z2	Z2	Z3	Z3	Z4	Z4	Z5	Z5
SPOOR PLATTING												
3. QRA	D											
Quadrant I	1	R	X	X	X	X	X	X	X	X	X	X
Quadrant II	2		X	X	X	X	X	X	X	X	X	X
Quadrant III	3		X	X	X	X	X	X	X	X	X	X
Quadrant IV	4		X	X	X	X	X	X	X	X	X	X
Quadrant I	1	L	X	X	X	X	X	X	X	X	X	X
Quadrant II	2		X	X	X	X	X	X	X	X	X	X
Quadrant III	3		X	X	X	X	X	X	X	X	X	X
Quadrant IV	4		X	X	X	X	X	X	X	X	X	X
SKE/PIC	E											
	1	S/P	P	P	P	P	P	P	P	P	P	P

Source: Created by author based on methodology from Tom Brown, Jr., *The Science and Art of Tracking: Nature's Path to Spiritual Discovery* (New York: Berkley Books, 1999), 144.

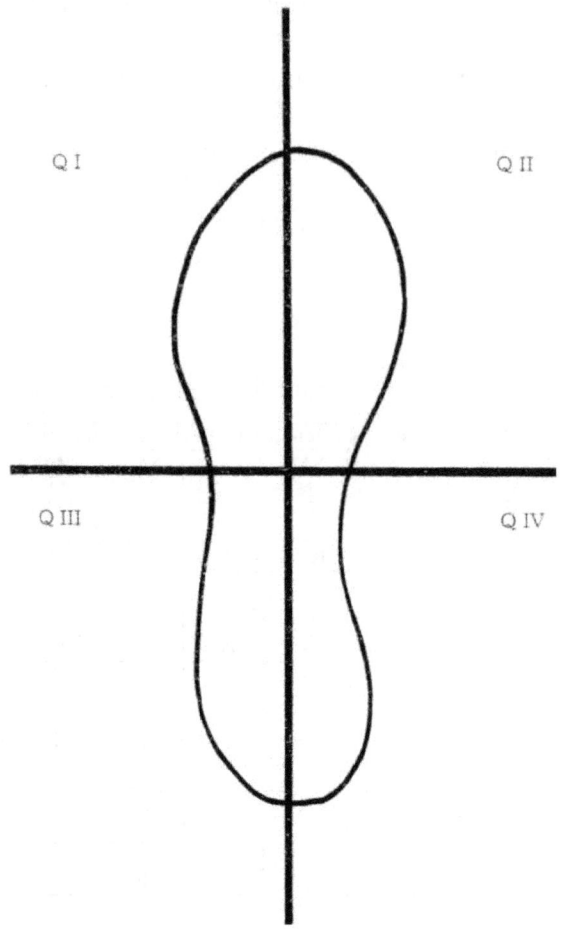

Figure 13. Quadrant Reference Analysis
Source: Tom Brown, Jr., *The Science and Art of Tracking: Nature's Path to Spiritual Discovery* (New York: Berkley Books, 1999), 144.

Foot Mapping Analysis II

Foot Mapping Analysis II (FMA-II) in table 7 is the grid locator of random

environmental encryptions that are branded into the manufacturers tread pattern through

wear interaction use with the environment. The analysis is the vertical axis and horizontal

axis intersecting the impression to compartmentalize the location of shoe print anomalies

added through trackmaker wear. These distinguishing marks are different from

manufacturing marks during creation of footwear. FMA-II targets the cuts, tears, chips, rock lodgments, wear degradation, etc. within the footprint that can be matched to the known shoe. In this case, the FMA-II is already established by the restraints and constraints of the experiment with the spoor analysis lanes and incident reconstruction area.

Table 7.   Foot Mapping Analysis II

OBSERVATION & SPOOR RECOGNITION (OSR)												
			SAL	IRA	SAL	IRA	SAL	IRA	SAL	IRA	SAL	IRA
LKT/LKS	A-V	OCISE	Z1	Z1	Z2	Z2	Z3	Z3	Z4	Z4	Z5	Z5

SPOOR PLATTING												
4. FMA2	F											
V.A. & H.A	1	RQ-I	X	X	X	X	X	X	X	X	X	X
A-H & 0-40	2	RQ-II	X	X	X	X	X	X	X	X	X	X
	3	RQ-III	X	X	X	X	X	X	X	X	X	X
	4	RQ-IV	X	X	X	X	X	X	X	X	X	X
	1	LQ-I	X	X	X	X	X	X	X	X	X	X
	2	LQ-II	X	X	X	X	X	X	X	X	X	X
	3	LQ-III	X	X	X	X	X	X	X	X	X	X
	4	LQ-IV	X	X	X	X	X	X	X	X	X	X

Source: Created by author based on methodology from Tom Brown, Jr., *The Science and Art of Tracking: Nature's Path to Spiritual Discovery* (New York: Berkley Books, 1999), 142-151.

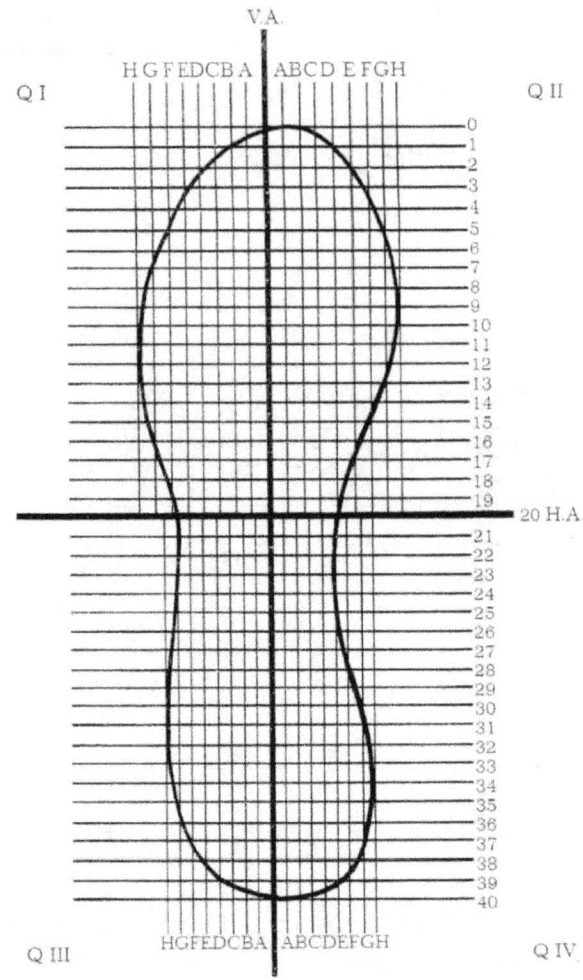

Figure 14. Foot Mapping Analysis II
Source: Tom Brown, Jr., *The Science and Art of Tracking: Nature's Path to Spiritual Discovery* (New York: Berkley Books, 1999), 148.

Ground Contact Points

The spoor analysis lanes and incident reconstruction area establishes the micro

class characteristics in size, shape, style, and pattern but the micro identifying

characteristics in the spoor analysis lanes is unessential to classification during this

experiment because the identity is already known. In the spoor analysis lanes, all

trackmakers Z1 through Z5 are using a normal walking gait, which registers all parts of

102

the foot: fore, mid, and hind. Because the substrate used the impact point, flex point and terminal point for normal walking gait it is distinct and pivot point is marginalized due to normal walking gait functions.

The ground contact point floor indicates normal walking by a wave, which measures either one-eighth or one-fourth inches respectively. No other ground contact points data will be seen on the floor because of the normal walking gait constraints within the spoor analysis lanes. The ground contact points wall (H22) shows a distinct cliff measuring one-eighth and the ground contact points horizon (H33) shows sporadic while the ground contact points horizon (H34) projects Z1, 3, and 5 drag of one-eighth to one-fourth and Z2 and 4 drag of one-fourth to three-eighths.

In incident reconstruction area, all situation trackmaker spoor-chain signatures were evaluated for identity using index comparative analysis because of non-visual observation. Index comparative analysis was 100 percent accurate in identification and the ground contact point indicator and fluctuating pressure release changes were due to trackmaker locomotor and psychomotor behavioral programming in primary movement patterns though complexity increased unilaterally throughout the gathering process of spoor evidence. Ground contact point's floor, wall, and horizon illustrated locomotor behavior for acceleration, pivoting, and stopping. Psychomotor behavior in the indicator pressure releases designated consistent human behavior identification of grief for situation 1, confusion/mild panic in situation 2, and calculation/thinking in situation 3.

Table 8.   Ground Contact Points

		OBSERVATION & SPOOR RECOGNITION (OSR)										
			SAL	IRA	SAL	IRA	SAL	IRA	SAL	IRA	SAL	IRA
LKT/LKS	A-V	OCISE	Z1	Z1	Z2	Z2	Z3	Z3	Z4	Z4	Z5	Z5

SPOOR PLATTING												
5. GCP												
MICC	G											
Design	1	Size	13.0		13.5		12.0		11.5		11.75	
	2	Shape										
	3	Style	Lug		Heel		Flat		Flat		Lug	
	4	Pattern	A1		A1		A1		A1		A1	
SD/FRA	H	RIGHT										
Fore foot	1	ff	Y	Y	Y	Y	Y	Y	Y	Y	Y	Y
Mid foot	2	mf	Y	Y	Y	Y	Y	Y	Y	Y	Y	Y
Hind foot	3	hf	Y	Y	Y	Y	Y	Y	Y	Y	Y	Y
Impact point	4	ip	Y	Y	Y	Y	Y	Y	Y	Y	Y	Y
Flex point	5	fp	Y	Y	Y	Y	Y	Y	Y	Y	Y	Y
Pivot point	6	pp	N	N	N	N	N	N	N	N	N	N
Terminal point	7	tp	Y	Y	Y	Y	Y	Y	Y	Y	Y	Y
	Floor	Right										
Wave	8	W	1/8	1/4	1/8	1/8	1/8	1/4	1/8	1/8	1/8	1/4
Double Wave	9	DW										
	Wall	Right										
Cliff	22	Cl	FL-1/8	FL-1/8	FL-1/8	FL-1/8	FL-1/8	FL-1/8	FL-1/8	FL-1/8	FL-1/8	FL-1/8
		1/8-full										
	Horizon	Right										
Toe Drag	33	1/8-full	Sprad	"	"	"	"	"	"	"	"	"
Heel Drag	34	1/8-full	1/8-1/4	1/8-1/4	1/4-3/8	1/4-3/8	1/8-1/4	1/8-1/4	1/4-3/8	1/4-3/8	1/8-1/4	1/8-1/4
SD/FRA	H	LEFT										
Fore foot	1	ff	Y	Y	Y	Y	Y	Y	Y	Y	Y	Y
Mid foot	2	mf	Y	Y	Y	Y	Y	Y	Y	Y	Y	Y
Hind foot	3	hf	Y	Y	Y	Y	Y	Y	Y	Y	Y	Y
Impact point	4	ip	Y	Y	Y	Y	Y	Y	Y	Y	Y	Y
Flex point	5	fp	Y	Y	Y	Y	Y	Y	Y	Y	Y	Y
Pivot point	6	pp	N	Y	N	Y	N	Y	N	Y	N	Y
Terminal point	7	tp	Y	Y	Y	Y	Y	Y	Y	Y	Y	Y
	Floor	Left										
Wave	8	W	1/8	1/4	1/8	1/8	1/8	1/4	1/8	1/8	1/8	1/4
Double Wave	9	DW										
	Wall	Left										
Cliff	22	Cl	FL-1/8	FL-1/8	FL-1/8	FL-1/8	FL-1/8	FL-1/8	FL-1/8	FL-1/8	FL-1/8	FL-1/8
		1/8-full										
	Horizon	Left										
Toe Drag	33	1/8-full	Sprad	"	"	"	"	"	"	"	"	"
Heel Drag	34	1/8-full	1/8-1/4	1/8-1/4	1/4-3/8	1/4-3/8	1/8-1/4	1/8-1/4	1/4-3/8	1/4-3/8	1/8-1/4	1/8-1/4

Source: Created by author.

Aerial Travel Points

The aerial travel points (ATP) suggest normal walking gait congruent in both the spoor analysis lanes and incident reconstruction area for all trackmakers. Z1, 3, and 5 have a slightly increased impact point gradient and terminal point gradient at 20 percent although this is not significant given the parameter variances for normal walking gait. The angle of increase in Z1, 3, and 5 is due to locomotor behavioral programming, which is conditioned. In Z2 and 4, the impact point gradient and terminal point gradient angle is lower at 10 percent suggesting a lower foot angle of exit and entry. The locomotor behavioral programming of Z2 and 4 is also conditioned.

The impact point gradient and terminal point gradient angles explain the apex of foot arc measurements for all trackmakers. However, the lower percentage angle the lower the apex of foot arc making measurement actualities acute thus the range of 1-2 in Z2 and 4. Above 20 percent the apex of foot arc is considered set allowing measurement precision to be a function of locomotor behavior energy fluctuation within each trackmaker. Thus, Z1, 3, and 5 movements through both spoor analysis lanes and incident reconstruction area is with more kinetic energy than is used by Z2 and 4.

In the incident reconstruction area, all trackmakers exhibited in situation 1 sorrow/grief indicator pressure releases through lowering the apex of foot arc reducing energy necessary to pick the feet up during locomotor behavior. A dragging sensation reverberated through the spoor-chain signature with erratic foot pitch and placement. The trail width analysis had no consistency with the baseline from the spoor analysis lanes, reaching in Z2, a maximum of twenty (20) inches. In situation 2, anxiety/mild panic showed confusion as to what to do; stops, turns, to look to see, which way to go.

Indecision to staying in place as to continuing was apparent. The aerial travel points at times were difficult to gage a consistent register deviation anomaly, switching as much as inches between step and stride analysis right to left. In situation 3, the spoor-chain signature was pure cognition. Calculating foot placement to minimize spoor-chain signature exposure as well as using route influencers to the physical environmental restrictions.

Table 9.   Aerial Travel Points

OBSERVATION & SPOOR RECOGNITION (OSR)												
			SAL	IRA	SAL	IRA	SAL	IRA	SAL	IRA	SAL	IRA
LKT/LKS	A-V	OCISE	Z1	Z1	Z2	Z2	Z3	Z3	Z4	Z4	Z5	Z5

GAIT PLATTING												

6. ATP	J	R/L										
Terminal Point Gradient	1	tpg	20%	5-30%	10%	5-30%	20%	5-30%	10%	5-30%	20%	5-30%
Apex of Foot Arc	2	apex	3	1-5	1-2	1-5	3	1-5	1-2	1-5	3	1-5
Impact Point Gradient	3	ipg	20%	5-30%	10%	5-30%	20%	5-30%	10%	5-30%	20%	5-30%
Terminal Point Gradient	1	tpg	20%	5-30%	10%	5-30%	20%	5-30%	10%	5-30%	20%	5-30%
Apex of Foot Arc	2	apex	3	1-5	1-2	1-5	3	1-5	1-2	1-5	3	1-5
Impact Point Gradient	3	ipg	20%	5-30%	10%	5-30%	20%	5-30%	10%	5-30%	20%	5-30%

Source: Created by author.

Stride and Step Analysis

The Stride and Step Analysis (SASA) measurements in table 10 within both the spoor analysis lanes and the incident reconstruction area for all trackmakers suggest unique locomotor and psychomotor behavioral programming. The spoor analysis lanes provide some constraints to assist in locomotor behavior baseline for Z1 through 5. In the incident reconstruction area, however, psychomotor behavioral programming begins to alter locomotor behavioral baselines in order to manipulated intent to environmental restraints. Thus, SASA changes in the incident reconstruction area to negotiate individualization.

No SASA baseline was possible in the incident reconstruction area because of the multiple psychomotor behavioral changes to the normal walking gait. The changes within the incident reconstruction area of Z1 through Z5 locomotor behavioral programming, establishes psychomotor behavior, which indicate commonality of intent within the mind of the sign story.

Table 10.   Stride and Step Analysis

OBSERVATION & SPOOR RECOGNITION (OSR)												
			SAL	IRA	SAL	IRA	SAL	IRA	SAL	IRA	SAL	IRA
LKT/LKS	A-V	OCISE	Z1	Z1	Z2	Z2	Z3	Z3	Z4	Z4	Z5	Z5

GAIT PLATTING												

7. SASA	K	R/L										
Stride Length	1	SL	56.75		53.0		58.0		42.5		44.5	
Step Interval	2	SI	21.0		22.75		23.75		13.0		16.0	
Stride Length	1	SL	57.0		52.0		56.5		42.0		45.0	
Step Interval	2	SI	22.0		19.5		23.0		12.5		15.75	

Source: Created by author.

Pitch Angle Analysis

In table 11 the registration of all trackmakers Z1 through Z5 are positive in both spoor analysis lanes and incident reconstruction area. The distance is measured in inches from the inner most point at the heel of registration in line with the direction of travel out to the tip of the print at the toe of registration; the higher the number, the greater the pitch. Thus, Z3 and Z5 use less pitch in locomotor behavior at 2 inches than does Z4 at 5 inches.

In the incident reconstruction area, there was not a change in PAA during normal walking gait even though the stride and step analysis was inconsistent. Thus, changing gait in the incident reconstruction area by each trackmaker to negotiate intent did not change the pitch. PAA did change, however, when locomotor behavior was influenced by psychomotor behavior outside the normal walking gait. When psychomotor behavior altered locomotor behavior thus changing the baseline, PAA could not be averaged for logging in the uniform scale format.

Table 11.   Pitch Angle Analysis

OBSERVATION & SPOOR RECOGNITION (OSR)												
			SAL	IRA	SAL	IRA	SAL	IRA	SAL	IRA	SAL	IRA
LKT/LKS	A-V	OCISE	Z1	Z1	Z2	Z2	Z3	Z3	Z4	Z4	Z5	Z5

GAIT PLATTING												
8. PAA	M	R/L										
Positive	1	pos	X	X	X	X	X	X	X	X	X	X
Negative	2	neg		X		X		X		X		X
Angle	3	ang		X		X		X		X		X
Distance/%	4	dis	4.0	4.0	4.5	4.5	2.0	2.0	5.0	5.0	2.0	2.0
Positive	1	pos	X	X	X	X	X	X	X	X	X	X
Negative	2	neg		X		X		X		X		X
Angle	3	ang		X		X		X		X		X
Distance/%	4	dis	4.0	4.0	4.5	4.5	2.0	2.0	5.0	5.0	2.0	2.0

Source: Created by author.

Trail Width Analysis

The Trail Width Analysis (TWA) in table 12 logged uniqueness in both straddle and trough for each trackmaker Z1 through Z5 during normal walking gait in both the spoor analysis lanes and the incident reconstruction area. In the incident reconstruction area, when the trackmakers locomotor behavior in normal walking gait was altered by psychomotor behavior during intent to negotiate environmental factors the trail width analysis at times became congested and unmeasureable.

Table 12.   Trail Width Analysis

OBSERVATION & SPOOR RECOGNITION (OSR)												
			SAL	IRA	SAL	IRA	SAL	IRA	SAL	IRA	SAL	IRA
LKT/LKS	A-V	OCISE	Z1	Z1	Z2	Z2	Z3	Z3	Z4	Z4	Z5	Z5
GAIT PLATTING												
9. TWA	N	R/L										
	1	stradl	1.75	1.75	1.5	1.5	-1.0/0	-1.0/0	4.5	4.5	2.5	2.5
	2	trough	12.0	12.0	12.0	12.0	7.0	7.0	14.0	14.0	12.5	12.5
	1	stradl	1.75	1.75	2.0	2.0	-1.0/0	-1.0/0	4.0	4.0	-1.0	-1.0
	2	trough	13.0	13.0	13.0	13.0	6.5	6.5	14.0	14.0	12.0	12.0

Source: Created by author.

Baseline

The indications within OCISE to this point of the algorithm show uniqueness that is distinguishable between trackmakers Z1 through Z5 during locomotor behavior in normal walking gait. The Baseline (BLE) for each trackmaker is capsulated in the measurements within the uniform scale format. This BLE is critical to see psychomotor behavior. Without locomotor behavior baseline from which the tracker can recognize register deviation anomalies, the psychomotor behavior is difficult at best to interpret clearly. Tracker knowledge that is comprehensive in understanding both universal locomotor behavior and psychomotor behavior generally as well as register deviation

109

anomaly alterations of locomotor and psychomotor behavior specifically will quickly glean understanding as to human behavior from the spoor-chain signature.

One of the key factors in any court of law is establishing a defendant's intent as to state of mind. Dispenza clearly states that the mind is the action arm of the brain. He further establishes the development of mind as neural programs that are controlled by the brain particularly the frontal lobe or neocortex. The neocortex is the thinking part of the brain, which interacts with the environment by commanding the body to engage the environment through movement. As described earlier in this thesis with the step-by-step example of how the brain, mind, and body interact to accomplish a task through primary movement patterns to an objective. To help illustrate Dispenza's theories further, in relation to the establishment of intent for courts of law; the following case studies will help to clarify BLE pragmatically and how the spoor-chain signature pinpoints register deviation anomalies, which clarifies human behavior.

Case study by a Game Warden Tracker for the State of Virginia

On January 2nd of this year, at approximately 6:00 PM, I was on patrol around Lake Anna in Louisa County, Virginia. As I approached a bridge that crosses over a channel where people often fish I noticed a vehicle that was parked in the woodbine rather than in the parking lot. I slowed down on the bridge so I could look both ways down the channel and I noticed two subjects that were approximately 150 yards down the channel. They spotted me at the same time and both immediately took off running into the woods to their rear. I crossed the bridge, pulled into the parking area and blocked their vehicle in with mine. I called for backup and began a foot pursuit after the two subjects. I was unable to see where the subjects had gone to and decided to go back to where I had last seen them and begin tracking them. I found there tracks and began to follow them. I noticed that there were more than two sets of tracks and again requested assistance and a K-9 to help locate the subjects. I thought that I had four sets of tracks and one set looked to be a juveniles or females because of the small size. I followed the tracks through the woods and could actually see a spot where one of the subjects had cleared a spot behind a tree so that he could watch for me without making noise. The tracks then continued on around the bend and down to the

110

water. The total distance of tracking was about 300 yards. I had backup within 30 minutes and the K-9 showed up shortly after that. It was right at dusk now and the K-9 was unable to pick up on a specific track. We decided to go back to where the subjects had entered the water and see if we could again pick up their tracks. As we approached the last known spoor we heard some noises and moved ahead to find five Cambodians crawling out of the water trying to sneak back to their vehicle. They were questioned as to why they ran and they said that they did not run, in fact they had never even known that I was there. There were 3 males and 2 females. They had been digging fresh mussels out of the lake in order to take back up north and sell in restaurants in Pennsylvania. They had taken over 1,700 pounds of mussels. A total of 11 charges were placed and one of the subjects was found to be an illegal alien. I was able to prove that they had run by taking pictures of the spot where I had seen them and was able to show the court the length of the strides and the sudden change in direction from when the subjects saw me and turned to run into the woods. I was also able to prove that they were trying to hide from me due to the sign that was found behind a tree where the subject had tried to observe me approach. I was also able to identify the subject that was hiding behind the tree because of his shoe prints. He later admitted that I was correct in my assumptions.[2]

This simple crime scene case illustrates the trackers knowledge in baseline observation and spoor recognition. He was able to understand psychomotor behavior through spoor platting, gait platting, and behavior platting. This gave the tracker the sign story comprehensibly, which allowed his expert testimony in court to substantiate his understanding of the spoor-chain signature and the quarry mental state. Had he not had been "spoor conscious," he would have not been able to have applied his deliberate collection of spoor evidence for the court. Most likely, they would have been successful in their flight from the area and no prosecution would have occurred.

Case study by an Alaska State Trooper Tracker

On 01-30-98, at about 2340 hours AST was advised that two persons were just observed attempting to break into the Tazlina store located near mile 111 of the Richardson Highway.

The conditions for this track were in darkness, ambient temperature about 8 F degrees above zero. A flashlight was used to observe the track evidence. Surface

111

conditions varied from hard packed ice and snow over pavement to deep, previously undisturbed snow two feet deep. The terrain was level.

The indicated direction of travel was north, somewhat parallel to the Richardson Highway. The initial start point of the track was the hard, snow packed surface of the driveway to the store.

The track in this location was characterized by faint sole impressions and heel marks. The tracks also had the effect of polishing the icy surface where contact was made, particularly the toe digs. The stride length exceeded forty inches, indicative of a person running. There were two distinct sets of tracks that maintained a parallel course to one another.

The track continued north across the parking lot and then proceeded onto a snow machine trail. The snow was not as firmly packed at this point, and distinguished characteristics of the individual sole patterns were observed.

The sole impressions observed were compared with those worn by the complainant, and they were found to be entirely different. One was a tennis shoe type style, and the other was a lug boot sole with a large gap between the heel and the ball.

At a point about one hundred yards down this path, both of the fleeing suspects had stopped running and turned to look back from where they had come. They then continued north on the path, although the stride had shortened considerably, as if the suspects were now jogging. Heel strikes were still deep, and the toe digs were forceful, causing the leading edge of the track to become distorted from slippage.

Near the intersection of Tazlina Terrace, which goes east from the Richardson Highway, the two tracks diverged. The suspect wearing the lug sole boots continued in a northerly direction and crossed suspect wearing the tennis shoes veered off the path, to the east through deep snow, and hid behind a large box that was situated about 30 feet from the snow machine trail. The place where the suspect laid down in the snow left a body size impression. The suspect did not remain at this location, however, as the track again crossed the deep snow to Tazlina Terrace, where it then went east and crossed the road in a diagonal manner.

The road surface was packed snow and ice on Tazlina Terrace. The track here was characterized by snow deposited from the sole of the tennis shoe onto the road surface. It continued to the north edge of the road until it entered the driveway that accesses the Ellis residence. The track proceeded into the driveway, and circled north around the house, before crossing the deep snow for about fifty feet. The track then entered the driveway of a residence that was part of the Copper River Housing Authority complex. It went north onto the street and was then

112

rejoined by the log boot sole, which had approached from the west. Both tracks then proceeded east, and terminated at the residence of Alice Craig.

The persons at the residence were identified as Ms. X, and two juveniles. They indicated that shortly after 2300 hours Mr. Y and Mr. Z had arrived at the residence, and that their pant legs were covered with snow. Mr. Y and Mr. Z had used the telephone, and shortly thereafter a red pickup truck arrived and they departed in it. Ms. X indicated that they had acted weird when they were here and she did not want them to stay.[3]

This simple crime scene case also illustrates the trackers knowledge in baseline observation and spoor recognition. They were able to substantiate the witness statements by what the spoor-chain signature showed for each of the quarry's and later made the arrests for attempted burglary. The spoor-chain signature gave the sign story, which showed quarry mental state. The AST trackers could establish locomotor behavior (running, jogging, turning, stopping, laying) and psychomotor behavior (cognition, affection, and conation) by the primary movement patterns and register deviation anomalies off the baseline. The trackers followed the spoor evidence step-by-step deliberately in order to maintain continuity and constancy within the BGS paradigm. Once the baseline is reputable, any register deviation anomaly becomes acutely observable even to a novice tracker as outlined in the first two case studies. In OCISE, the behavior platting portion of the algorithm starts with locomotor behavioral programming and moves to the advanced analysis of human behavior, which only an expert or master tracker can glean from the spoor-chain signature.

In table 13, the body state as to locomotor behavior is recognized. Both in the spoor analysis lanes and within the incident reconstruction area, all trackmakers Z1 through Z5 showed through the spoor-chain signature in p1, a 1a, which is the normal walking gait. In the incident reconstruction area, however, tasks were given, which required the trackmakers to engage the environment with the neocortex, which showed intent by the thinking part of the brain. To accomplish the task each trackmaker had to register deviation anomaly off the baseline in order to be successful in accomplishing the task. These register deviation anomalies employed the locomotor behavior (fluctuating pressure releases) for turning, stopping, sitting, and standing. Also, noticed in the locomotor behavioral programming was kneeling, bracing the body with the hands on the ground on all fours, and lying on the ground.

In the incident reconstruction area, the gait-footfall sequence enablers were established for all trackmakers. Z1 through Z5 were all fit comparable to the task requirements. They all were stable with no injuries or wounds that would affect the results of the tasks. Therefore, blood spoor was not evaluated. Each trackmakers resistance was only their respective body weights. Z1, 2, 3, and 5 were adult males and Z4 was an adult female. There was no assistance need from the head, arms, and torso for stability in locomotor behavior except during one task to pull something out at a low level where assistance of the arm and hand was necessary to brace the body at such a low angle.

Table 13. Locomotor Behavioral Traits and Gait Footfall Sequences Effectors/Enablers

BASELINE ESTABLISHMENT (BLE)												
			SAL	IRA	SAL	IRA	SAL	IRA	SAL	IRA	SAL	IRA
LKT/LKS	A-V	OCISE	Z1	Z1	Z2	Z2	Z3	Z3	Z4	Z4	Z5	Z5

BEHAVIOR PLATTING												
LMBP Body State	P											
Walking	1		1a	1a	1a	1a	1a	1a	1a	1a	1a	1a
Running	2		N	N	N	N	N	N	N	N	N	N
Turning	3		Y	Y	Y	Y	Y	Y	Y	Y	Y	Y
Slipping/ Falling	4		N	N	N	N	N	N	N	N	N	N
Sitting/Standing	5		Y	Y	Y	Y	Y	Y	Y	Y	Y	Y
Stopping	6		Y	Y	Y	Y	Y	Y	Y	Y	Y	Y
Crawling/ Rolling	7		N	N	N	N	N	N	N	N	N	N

GFS Effectors/ Enablers	S											
Fit/Unfit	1		F	F	F	F	F	F	F	F	F	F
Stable/Unstable	2		S	S	S	S	S	S	S	S	S	S
Injured/ Structure Problems	3		NEG	NEG	NEG	NEG	NEG	NEG	NEG	NEG	NEG	NEG
Wounded/ Blood Spoor	4		NE	NE	NE	NE	NE	NE	NE	NE	NE	NE
Heavy/Light Load	5		NEG	NEG	NEG	NEG	NEG	NEG	NEG	NEG	NEG	NEG
Male/Female	6		M	M	M	M	M	M	F	M	M	M
Child/Teen/ Adult	7		A	A	A	A	A	A	A	A	A	A
HAT Assist for stability	8		NEG	NEG	NEG	NEG	NEG	NEG	NEG	NEG	NEG	NEG

Source: Created by author.

Register Deviation Anomalies

In table 14, the Register Deviation Anomalies (RDA) within the spoor analysis lanes was controlled by locomotor behavior constraints in order to get a near perfect baseline of each trackmaker without much macro class characteristic distortions. In the incident reconstruction area, the psychomotor behavior was the focus of the experiment, and therefore, each trackmaker could negotiate the tasks at will so that register deviation anomalies were natural extensions of the senses of the body in response to the environment as to establish macro-identifying characteristic register deviation anomalies for each trackmaker. The macro-class characteristic register deviation anomalies contained no weather-imposed distortions. Self-imposed distortions were not evaluated; however, they are always present when human locomotion encounters transfer of foot weight acceptance during the gait cycle. The substrate in any environment is unpredictable in weight acceptance and therefore alters the way locomotor behavior affects the spoor-chain signature during locomotion.

In the spoor analysis lanes and the incident reconstruction area, both mechanical imposed distortions and landscape-imposed distortions were present. All trackmakers Z1 through Z5 were wearing shoes, which influenced the impact point and terminal point along the spoor-chain signature. The locomotor behavior embraced these distortions, which assisted the experiments in both the spoor analysis lanes and the incident reconstruction area comprehensibly leading to macro-identifying characteristic analysis formulations off each trackmakers locomotor behavior to solidify the psychomotor behavior of all trackmakers.

Macro identifying characteristics is the internal adjustments that a trackmaker must make based on the five senses as interaction with the environment sends signals to the brain for action by the mind of each trackmaker. Both Psychomotor anomaly and Loco-motor anomaly of all five trackmakers were recognized in the incident reconstruction area. The findings of psychomotor and locomotor anomalies in any spoor-chain signature require the expansion of behavioral platting into the eight adaptive traits, which gives clarity to the anomaly and introduces the human behaviors to the tracker.

116

Table 14.   Register Deviation Anomalies

BASELINE ESTABLISHMENT (BLE)												
			SAL	IRA	SAL	IRA	SAL	IRA	SAL	IRA	SAL	IRA
LKT/LKS	A-V	OCISE	Z1	Z1	Z2	Z2	Z3	Z3	Z4	Z4	Z5	Z5

BEHAVIOR PLATTING												
10. RDA	T											
MACC	1	WID	N	N	N	N	N	N	N	N	N	N
(external))	2	SID	NE	NE	NE	NE	NE	NE	NE	NE	NE	NE
	3	MID	Y	Y	Y	Y	Y	Y	Y	Y	Y	Y
	4	LID	Y	Y	Y	Y	Y	Y	Y	Y	Y	Y
MAIC	5	PMA	Y	Y	Y	Y	Y	Y	Y	Y	Y	Y
(internal)	6	LMA	Y	Y	Y	Y	Y	Y	Y	Y	Y	Y

Source: Created by author.

Eight Adaptive Traits

Tables 15 and 16 work together and outline the course of action, which the neocortex framed in the mind of each of the trackmakers Z1 through Z5 respectively for each task within the incident reconstruction area. In U1, 2, and 3 all five trackmakers displayed neocortical framing within the manifest cognitive/intuitive trait, manifest volitional trait, and manifest imperturbable-mind/steadfast-mind trait by performing evaluation and making decisions based on personal intra-cranial perceptions of the problem to be solved. This executive functioning within the three brain-bound traits was manifest through locomotor behavioral programming with psychomotor behavior programming register deviation anomalies showing initiative, forbearance, and the ability to override the locomotor behavior of inadequate ability on task to try novel ideas formulated in the neocortex.

In U4, 5, 6, 7, and 8 all five trackmakers transferred U 1, 2, and 3 executive functioning within the neocortex to implementation through the two body-bound traits and three action-bound trait. All five trackmakers displayed the ability in U9 to perform

117

locomotor, which was psychomotor behavior based on past levels of experience ranging from very familiar to minimally familiar. Only Z1, 3, and 5 were able to employ U10 idiosyncratic motor behavior alterations, which were system free from the system bound psychomotor behavior. In U12, all trackmakers displayed midbrain functioning when confronted with interruption through environmental stimulus, which invoked surprise/fear in fight or flight.

Table 15. Eight Adaptive Traits

BASELINE ESTABLISHMENT (BLE)												
			SAL	IRA	SAL	IRA	SAL	IRA	SAL	IRA	SAL	IRA
LKT/LKS	A-V	OCISE	Z1	Z1	Z2	Z2	Z3	Z3	Z4	Z4	Z5	Z5

BEHAVIOR PLATTING												
11. EAT	U	IAT/ MAT										
Three BRBT	1	MCIT	X	X	X	X	X	X	X	X	X	X
	2	MVT	X	X	X	X	X	X	X	X	X	X
	3	MIST	X	X	X	X	X	X	X	X	X	X
Two BOBT	4	MABT	X	X	X	X	X	X	X	X	X	X
	5	MRVT	X	X	X	X	X	X	X	X	X	X
Three ACBT	6	MOPT	X	X	X	X	X	X	X	X	X	X
	7	MFYT	X	X	X	X	X	X	X	X	X	X
	8	MST	X	X	X	X	X	X	X	X	X	X
PMBP	9	Sys Bnd	X	X	X	X	X	X	X	X	X	X
IMBP	10	Sys Free	X	X	X	X	X	X	X	X	X	X
GAS/GIS w/Relationship	11	SNS/ PSNS	X	X	X	X	X	X	X	X	X	X

Source: Created by author based on methodology from Richard Hayes, *Paleolithic Adaptive Traits and the Fighting Man* (Sedona, AZ: International Hoplology Society, 1998).

Transcendence Synergy of the Manifest Adaptive Traits

Table 15 set the stage for the details in table 16. V1, 2, and 3, thus establishing the locomotor behavior programming of the Transcendence Synergy of the Manifest Adaptive Traits (TS/MAT) of each trackmaker. Where there is a register deviation anomaly off the locomotor behavior, a psychomotor behavior through interaction with stimulus within the environment helps the performance and outcome by delimiting initiative or unlimiting initiative in the lowland and midland areas. At the upland area, the initiative either withheld action or took action based on the external stimulus. Thus, V2 proved W2 through eight by each trackmaker exhibiting non-grasping persona or intra-psychic factors that controlled locomotor behavior output through executive functioning, implementation, and performance and outcome.

Table 16. Transcendence Synergy of the Manifest Adaptive Traits

			BASELINE ESTABLISHMENT (BLE)									
			SAL	IRA	SAL	IRA	SAL	IRA	SAL	IRA	SAL	IRA
LKT/LKS	A-V	OCISE	Z1	Z1	Z2	Z2	Z3	Z3	Z4	Z4	Z5	Z5

BEHAVIOR PLATTING												
12. TS/MAT	V											
Del/Unl/Mel	1		NE	X	NE	X	NE	X	NE	X	NE	X
GP/NGP	2		NE	X	NE	X	NE	X	NE	X	NE	X
L/M/U P&O	3		NE	X	NE	X	NE	X	NE	X	NE	X
PMBP (Mental State)	W											
Nervous/ Anxious	1		N	N	N	N	N	N	N	N	N	N
Quiet/ Thoughtful	2		Y	Y	Y	Y	Y	Y	Y	Y	Y	Y
Thinking/ Reasoning/ Planning	3		N	Y	N	Y	N	Y	N	Y	N	Y
Panic/Fear (Phobia)	4		N	Y	N	Y	N	Y	N	Y	N	Y
Hurrying/Haste	5		N	Y	N	Y	N	Y	N	Y	N	Y
Calm/Coolness/ Total Control	6		Y	Y	Y	Y	Y	Y	Y	Y	Y	Y
Out of Control/ Unthinking	7		N	Y	N	Y	N	Y	N	Y	N	Y
Aggressive/ Passive	8		N	N	N	N	N	N	N	N	N	N
SCS Effectors/ Enablers Culture/ Subculture Programming	X											
Occupational Traits	3		Y	Y	N	N	Y	Y	N	N	Y	Y
Trained/ Untrain Traits	5		T	T	UT	UT	T	T	UT	UT	T	T

Source: Created by author based on methodology from Richard Hayes, *Paleolithic Adaptive Traits and the Fighting Man* (Sedona, AZ: International Hoplology Society, 1998).

Human Behavior Identification

The use of OCISE in both the spoor analysis lanes and incident reconstruction area, in detail, establishes the manifestation within the spoor-chain signature the human behavior of each trackmaker.

120

<u>Summary</u>

This chapter applied the OCISE algorithm as criteria methodology, based on the science of linearity and laws of motion within the BGS paradigm, to assess the extent to which spoor evidence will elucidate human behavior thus addressing the primary and secondary research questions. With the collection of spoor evidence data, a breakdown of the collected data from the OCISE algorithm was recorded in the uniform scale format to answer whether human behavior can be gleaned from the spoorological record. The next chapter highlights the merits of the BGS paradigm (law of linearity and motion) and OCISE algorithm while giving due acknowledgement to the limitations inherent in the implementation of this research. The chapter ends with a discussion of recommendations of further research.

---

[1]Joel Hardin and Matt Condon, 256.

[2]Kevin Berger, e-mail to Michael Hull, 7 July 1999.

[3]Craig Allen, Alaska State Trooper Tracker, e-mail to author, 31 January 1998.

CHAPTER 5

CONCLUSIONS AND RECOMMENDATIONS

Having a non-tracker decide whether tracking will work is like having a painter decide whether or not surgery is required.

— Ab Taylor and Donald C. Cooper,
Fundamentals of Mantracking: The Step-By-Step Method

The last chapter presented and analyzed the spoor analysis lane and incident reconstruction area data, based on the OCISE algorithm, to determine whether the data collected would glean human behavior from the spoorological record. Several case studies were examined where locomotor and psychomotor behavioral was determined during actual tracking cases. This chapter highlights the merits of the OCISE algorithm while giving due acknowledgement to the limitations inherent in the implementation of this research. The chapter ends with a discussion of recommendations of further research.

Overview

This thesis explored the subject of human behavioral influences on spoor evidence and the intrinsic OCISE modeling based on primary and secondary movement patterns that are left as signature "sign" stories imprinted in the substrate. The sign story[1] then projects walking gait encryptions that can only be analyzed properly by the expertise of trackers. The design of the OCISE modeling-based representation for overall decryption of human movement to facilitate recognition, classification, and interpretation of people and their behaviors by their footfall sequences was evaluated. The OCISE method encompassed a "totality of the circumstances" approach to answering the interrelated questions: Is human behavior recognizable in the gait-footfall sequence and spoor-chain signature? In addition, how much locomotor and psychomotor information is contained in the collection of spoor evidence? The spoor evidence data was tested in this study by direct visual observation within spoor analysis lanes and confirmed by non-visual observation within an incident reconstruction area to validate collected research data based on the BGS Paradigm.

Many cases have implemented the use of spoor evidence related to footwear impressions and other collateral footprint testimony. Spoor evidence is located at every crime scene because wherever people walk they leave spoor-chain signatures. The problem is most investigators and crime scene technicians only see the obvious spoor. They fail to see the more numerous spoor, which is not looked for, and thus cannot be seen. The fact that most investigators and technicians will collect the obvious spoor is recognition of spoor evidence value to the court. The key to maximizing the effectiveness of this research would be to mandate the adoption of the OCISE for the purpose of instruction and implementation to see more spoor, which is present at each crime scene. By accessing the more numerous spoor through OCISE allows the investigator or technician to recognize human behavior, which gives direct access to the mind of the quarry.

Tracking Experiments

Two overall experiments were performed during this thesis. The first was by applying OCISE methodology to five separate spoor-chain signatures within spoor analysis lanes through direct visual observation. This experiment set the baseline for each of the trackmakers individually through spoor, gait, and behavior platting while evaluating trackmaker cognition. The second was by applying OCISE methodology to

123

five separate spoor-chain signatures, in three situations within incident reconstruction area through non-visual observation. This experiment reconfirmed the baseline for each trackmaker through spoor, gait, and behavior platting while evaluating trackmaker cognition, affection, and conation. The spoor analysis lane experiment was very straightforward in data collection because of tracker direct visual observation, which connected the spoor-chain signature to the observable gait-footfall sequence and human behavior identification. The incident reconstruction experiment introduced the possibility of validity error in human behavior identification due to removal of tracker direct visual observation during trackmaker primary movement pattern locomotor and psychomotor behavior.

Although the scale of the two experiments performed were not large enough to get a thorough quantitative view of OCISE values for determining statistical accuracies, it is possible to gain some relevant indications from these spoor calculations for qualitative analysis. One obvious observation is that the results from both experiments gleaned reliability as to both baseline and human behavior identification when the tracker is experienced in all modalities within OCISE. The two experiments indicated the strength of OCISE as sound methodology to view human behavior identification within a spoor-chain signature. It is however important to remember that when analyzing the uniform scale format of the raw data from OCISE, it is critical that the tracker is oriented the same way as the collector of the data (if not the same person) by seeing the evidence of the tracker as taken from the substrate. This is more important for human behavior identification than it is for baseline. The strongest analysis is, therefore, by the tracker who visually saw the spoor-chain signature first-hand rather than the tracker who is evaluating the data within the uniform scale format through non-visual observation of the data alone.

Depending on the level of certainty that is necessary, having an experienced tracker collect the data from a spoor-chain signature through direct visual observation requires a crime scene methodology switch by the current scientific and professional communities because the tracker would need to be present as a first responder to any scene. However, OCISE might still be sufficient for baseline establishment and human behavior identification should the many spoor destroyers continue to contaminate the physical evidence of tracks by trampling the substrate while collecting other evidence, such as fingerprints, deemed as valuable to the forensic sciences if an experience tracker was even called to every crime scene no matter the timing.

Further Research

This thesis has opened a new door in the field of forensic spoorology. As the task of developing the OCISE methodology from the BGS paradigm began to unfold, and as OCISE was performed, new problems and possibilities for further research were discovered. What follows is a list of the possible next steps of research that should follow this thesis.

Complexity of Substrates and Terrain

As this research was limited to a substrate that was flat with no vegetation. The next step is to begin to increase the substrate difficulty through adding in various ground hardness types and ground vegetation types from different locations geographically.

Next, look at elevation changes and the impact of these changes on primary movement patterns from locomotor and psychomotor behavioral programming.

Time and Distance Variances

Gradually increase the distance of the spoor-chain signature to isolate longevity of locomotor behavior to see the impact this may have on psychomotor behavior.

Observation and Spoor Recognition Deterioration

Studies into how the eyes deteriorate through constant contact with the spoor-chain signature may yield strategies to improve visual stamina and accuracy in spoor evidence recovery.

Human Behavior Identification Direct Visual Observation

Breaking down all the psychomotor anomalies numerically for direct visual observation analysis will glean specifics into ground contact point indicator pressure releases and fluctuating pressure releases so they can be recorded specifically, instead of generally.

Rate of Error Mitigation through Constancy

One of the BGS paradigm footings for its construction was the law of linearity, which produces constancy as one of the prongs. Does constancy reduce rate of error in accuracy, reliability, and validity?

Forensics Yield of Other Crime Scene Evidence

When tracking is used as first response methodology to a crime scene is there an increased yield of other forensic evidence because all spoor evidence is found, thus leading directly to those places that the suspect trackmaker actually was.

Victim, Suspect, Witness Spoor Discrimination

A detailed longitudinal analysis of the baseline and human behavior identification of crime scene victims, suspect, and witnesses would assist in quantitative and qualitative data.

Universal Tracker Certification Standardization for Forensic Spoorology

OCISE must be employed by an experienced tracker for greatest reliability and validity for the courts. What testing measures would standardize tracker skill level, thus enhancing OCISE usability?

Conclusions

The greatest challenge in applying the OCISE algorithm to a crime scene lies in the skill level of the investigative or police tracker. Thomas Kuhn, in *The Structure of Scientific Revolutions*, capsulated the structure of new paradigms as being difficult to adopt.[2] Further research must be conducted into this difficulty and why paradigms are so suspect by trackers, in general, and by scientists, more specifically.

The results of the OCISE algorithm have been discussed in the previous chapter. The focus has been to show the utility of the algorithm for establishing scientific reliability for the admissibility of human behavior designation based on a well-

127

established spoor-chain signature. The next step in the implementation of this OCISE algorithm for investigative and police trackers would be for the jurisdictional courts at all levels to base tracking testimony reliability on this algorithm. To make this research valuable, subsequent implementation and training should be conducted within all police and forensic jurisdictions that are responsible for responding to various crime scenes where spoor evidence is missed, destroyed, and not collected. This would allow procedures and focus instruction at the field level to familiarize all evidence gatherers to be what master tracker Joel Hardin calls, "Track Aware," which would therefore maximize spoor's evidentiary value for the courts because "that which is looked for will be seen."

This thesis started by asking: does the spoor evidence collected through observation, classification, and interpretation of spoorological research establish the scientific accuracy requirements for court viability and reliability; and, can this OCISE methodology include or exclude suspects, and in turn, help the criminal justice system work more effectively; and, is human behavior recognizable in the spoor-chain signature? As the thesis moved forward, it was clear that these three questions were much more difficult questions than first thought, and a wide area of new questions were identified as can be ascertained by the further research proposals addressed earlier in this chapter. This research thesis has shown that human behavior is identifiable in both gait-footfall sequences and spoor-chain signatures. Also, all locomotor and psychomotor behavioral programming information is contained in the spoor evidence (GCP IPR and FPR) if it is observed, classified, and interpreted by experienced trackers and does therefore establish

128

scientific accuracy requirements needed for court viability and reliability with trackmaker inclusion and exclusion precision.

The experiments performed in this research shows that it is possible to collect reliable and valid baseline and human behavior identification data from a trackmakers spoor-chain signature. Through application of OCISE algorithm methodology based on the BGS paradigm of linearity and laws of motion, the sign story of each trackmaker is present at every crime scene. However, for tracking methods to achieve scientific acceptance as a forensic enabler, more research work is needed and more tracker education and training must be conducted to anyone who will learn.

One thing stands resolute through this thesis research: Just as fingerprint evidence isolates human identity to the person that left it, footprint evidence registers human behavior from the mind of the quarry that left it. The only question that could not be resolved by this thesis was; which tracker will get to the spoor evidence to read it, before the sign story is destroyed; by those who believe that it is impossible to see it, until they are trained to see it and read it, so that they will then believe it? "To the quarry lies the tracker and to the tracker lies the trail.[3] There is always a trail."[4] "Remember. . . . Remember. . . . Wherever people walk they leave sign."[5]

---

[1]Hardin and Condon, 256.

[2]Thomas Kuhn, *The Structure of Scientific Revolutions,* 2d ed. (Chicago, IL: University of Chicago Press, 1962), 168-170.

[3]Cunningham, 481.

[4]Ibid.

[5]Hardin and Condon.

APPENDIX A

ADDITIONAL ACRONYM LIST

Observation (Locating, Following)

Classification (Collecting, Recording)

Interpretation (Examining, Analyzing)

AGE	Age of Spoor
AS	Aerial Spoor
ASE	Apex Step Estimate
AT	Active Trail
BkTM	Back Trailing Method
BRKT	Bracketing
BTM	Bound Trailing Method
COE/E	Cone of Entry/Exit
CTM	Cut Trailing Method
DOS	Direction of Spoor
DOT	Direction of Trail
DTM	Deliberate Trailing Method
DyTM	Dynamic Trailing Method
EVCC	Evader Class Characteristics
FHD	Foot Horizontal Drag
FIRM	Foot Impression Reference Matrix
FMA	Foot Measurement Analysis
FOS	Forward Observable Spoor
FPC	Footprint Classification

FRA	Footroll Analysis
FVL	Foot Vertical Lift
FWC	Footwear Classification
GCP	Ground Contact Points
GHT	Ground Hardness Type
GP	Gait Platting
GS	Ground Spoor
GVT	Ground Vegetation Type
ICA	Index Comparative Analysis
ISPT	International Society of Professional Trackers
IT	Inactive Trail
KNQ	Knowledge of Quarry
LITDSC	Litter Discipline Classification
LSP	Linear Search Pattern
LTP	Lost Trail Procedures
MSZ	Mechanical Stride Zero
QPC	Quarry Physical Condition
QRA	Quadrant Reference Analysis
RDA	Register Deviation Anomaly
RSP	Rota Search Pattern
SLA	Sun Light Angle
SMD	Stick Measuring Device "Smiddy"
SPT	Straddle, Pitch, Trough
STA	Slope Topography Analysis
TDI	Time Distance Interval

TEC	Track Erosion Computation
TLE	Trails Leading Edge
TRA	Terrain Registration Area(s)
TS	Tertiary Spoor
TSE	Time Shadow Effect
TSP	Transverse Search Pattern
VHP	Vegetation Healing Process
VTP	Variable Trail Pattern

APPENDIX B

COMPLETE USF SYNCH MATRIX

USF SYNCH MATRIX													

Researcher(s):			Date:		Location:								

OBSERVATION & SPOOR RECOGNITION (OSR)													
LKT/LKS	A-X	OCISE	SAL Z1	IRA Z1	SAL Z2	IRA Z2	SAL Z3	IRA Z3	SAL Z4	IRA Z4	SAL Z5	IRA Z5	
SPOOR PLATTING													
1. COS	A												
	1												
	2												
	3												
	4												
	5												
2. FMA	B												
	1	R											
	2												
	3												
	4												
	5												
	1	L											
	2												
	3												
	4												
	5												
FIRM	C												
	1	Hgt											
	2	Wgt											
	3	MSZ											
3. QRA	D												
	1	R											
	2												
	3												
	4												
	1	L											
	2												
	3												
	4												
SKE/PIC	E												
	1	Sketch											
	2	Photo											
4. FMA2	F												
V.A. & H.A	1	RQ-I											
A-H & 0-40	2	RQ-II											
	3	RQ-III											
	4	RQ-IV											

	1	LQ-I											
	2	4LQ-II											
	3	4LQ-III											
	4	4LQ-IV											
5. GCP													
MICC	G												
Design	1	Size											
	2	Shape											
	3	Style											
	4	Pattern											
MIIC	1	Size											
Damage	2	Shape											
	3	Orient											
	4	Pos											
SD/FRA	H	R											
Eighth App	1	ff											
	2	mf											
	3	hf											
	4	ip											
	5	fp											
	6	pp											
	7	tp											
	Floor													
	8	W											
	9	DW											
	10	D											
		1/8-full											
	11	R/D-s											
	12	R/D-f											
		1/8-full											
	13	R/D-c											
		1/8-full											
	14	R/Di											
		1/8-full											
	15	R/Di-s											
	16	R/Di-f											
		1/8-full											
	17	R/Di-c											
		1/8-full											
	18	R/WO											
		1/8-full											
	19	R/WO-f											
		1/8-full											
	20	R/WO-c											
		1/8-full											
	21	R/Ex-off											
		1/8-full											

	Wall											
	22	R/Cl										
		1/8-full										
	23	R/Ri										
		1/8-full										
	24	R/Pe										
		1/8-full										
	25	R/Cr										
		1/8-full										
	26	R/Cr-Cr										
		1/8-full										
	27	R/Ca										
		1/8-full										
	28	R/Ca-in										
		1/8-full										
	Horizon											
	29	R/Pl										
		1/8-full										
	30	R/Pl-f										
		1/8-full										
	31	R/Pl-c										
		1/8-full										
	32	R/Pl-ex										
		1/8-full										
SD/FRA	H	L										
Eighth App	1	ff										
	2	mf										
	3	hf										
	4	ip										
	5	fp										
	6	pp										
	7	tp										
	Floor											
	8	W										
	9	DW										
	10	D										
		1/8-full										
	11	D-s										
	12	D-f										
		1/8-full										
	13	D-c										
		1/8-full										
	14	Di										
		1/8-full										
	15	Di-s										
	16	Di-f										
		1/8-full										
	17	Di-c										
		1/8-full										
	18	WO										
		1/8-full										

	19	WO-f									
		1/8-full									
	20	WO-c									
		1/8-full									
	21	Ex-off									
		1/8-full									
	Wall										
	22	Cl									
		1/8-full									
	23	Ri									
		1/8-full									
	24	Pe									
		1/8-full									
	25	Cr									
		1/8-full									
	26	Cr-Cr									
		1/8-full									
	27	Ca									
		1/8-full									
	28	Ca-in									
		1/8-full									
	Horizon										
	29	Pl									
		1/8-full									
	30	Pl-f									
		1/8-full									
	31	Pl-c									
		1/8-full									
	32	Pl-ex									
		1/8-full									
		GAIT PLATTING									
6. ATP	J	R/L									
	1	tpg									
	2	apex									
	3	ipg									
	1	tpg									
	2	apex									
	3	ipg									
7. SASA	K	R/L									
	1	SM									
	2	sm									
	1	SM									
	2	sm									
8. PAA	M	R/L									
	1	pos									
	2	neg									
	3	ang									
	4	dis									
	1	pos									
	2	neg									
	3	ang									
	4	dis									

9. TWA	N	R/L										
	1	stradl										
	2	trough										
	1	stradl										
	2	trough										
BASELINE ESTABLISHMENT (BLE)												
BEHAVIOR PLATTING												
LMBP Body State	P											
Walking	1											
Running	2											
Turning	3											
Slipping/ Falling	4											
Sitting/Standing	5											
Stopping	6											
Crawling/ Rolling	7											
GFS Effectors/ Enablers	S											
Fit/Unfit	1											
Stable/Unstable	2											
Injured/ Structure Problems	3											
Wounded/ Blood Spoor	4											
Heavy/Light Load	5											
Male/Female	6											
Child/Teen/ Adult	7											
HAT Assist for stability	8											
RDA	T											
MACC	1	WID										
(external)	2	SID										
	3	MID										
	4	LID										
MAIC	5	PMA										
(internal)	6	IMA										
	7	LMA										
10. VIII-AT	U	IAT/ MAT										
Three Brn-BT	1	IM/ SMT										
	2	C/IT										
	3	VT										
Five Bod-BT	4	OPT										
	5	AT										
	6	RVT										
	7	FYT										
	8	ST										
PAMP	9	Sys Bnd										
IBMP	10	Sys Free										

GAS/GIS w/Relationship	11	SNS/ PSNS											
12. TS/MAT	V												
Del/Unl/Mel	1												
GP/NGP	2												
L/M/U P&O	3												

HUMAN BEHAVIOR IDENTIFICATION (HBI)

PMBP (Mental State)	W												
Nervous/ Anxious	1												
Quiet/ Thoughtful	2												
Thinking/ Reasoning/ Planning	3												
Panic/Fear (Phobia)	4												
Hurrying/Haste	5												
Calm/Coolness/ Total Control	6												
Out of Control/ Unthinking	7												
Aggressive/ Passive	8												
SCS Effectors/ Enablers Culture/ Subculture Programming	X												
Urban/Rural Traits	1												
Nationality Traits	2												
Occupational Traits	3												
Criminal/Non-Criminal Traits	4												
Trained/ Untrain Traits	5												

NOTES/COMMENTS

Results													

APPENDIX C

TRACKER CRIME SCENE KIT USED FOR FIELD EXPERIMENTS

<u>CASE</u>
USMS Scout Tracker Logo
Carrying Case

<u>BACK POUCH</u>
Tape Measure 6ft & 25ft
Wire Flagging
"O" Rings

<u>TOP POUCH</u>
Gloves (Rubber)
Ziploc Bags (S,M,L)
35 Film Canisters
Plastic Storage Bottles
Navigational Compass
Whistle

<u>UNDER TOP FLAP</u>
Stick Measuring Device
  550 Cord
  100 Mile Tape

<u>UNDER FLAP-OUTER MOST POCKETS</u>
Markers (Black permanent)
Pencil (Mechanical & Regular & Colored)
Pen (Ink-Colored)
Business Cards

<u>VERTICAL ZIPPED POCKET</u>
Nails (16 Penny)
Popsicle Sticks
Small Tent Pegs

<u>OUTER MOST HORIZONTAL ZIPPED POCKET</u>
Eraser
Lead (Spare)
Ruler (6" & 12")
Triangle (45 & 60 Deg.)
Protractor (4")
Compass (4")
Crime Scene Template

MIDDLE HORIZONTAL ZIPPED POCKET
Marking Chalk (Multi)
Acetone Free Polish Remover
Lemon Nail Polish Remover
Iso Alcohol

INNER HORIZONTAL ZIPPED POCKET
Flagging Tape (Two rolls)
Perimeter Marking String (Colored)
Spoor Cards
Human Body Template
Pad (Graph Paper)
USMS-USF
USMS-Crime Scene Primer (W, U, A)
"Futprinter" Device
        Pencil (Grease)
Animal Pictures (Footprint & Bio)
Crime Scene Checklist
Tracker Field Notebook
Pad (Sticky)
Notepad (Small)

ON BODY
Wide-brim bush hat
Leather Gloves

ADDITIONAL ITEMS
Camera Digital

APPENDIX D

RECOMMENDED TRACKER CERTIFICATION STANDARDS

<u>Tracker Certification</u>

Tracker certification requires that applicants meet specified requirements prior to beginning the testing process, which consists of a written and practical test.

<u>Tracker Certification Requirements</u>

Qualifications, requirements and application procedures for certification are subject to revision by the Board.

General

An applicant for certification must be of good moral character, high integrity, good repute and must possess a high ethical and professional standing.

An applicant for certification must be primarily employed in the field of law enforcement or forensics and whose duties include the examination of spoor evidence.

Training Requirements

Applicants must have satisfactorily completed a tracker training program in the examination of spoor evidence to include, but not limited to, terminology, OCISE algorithm or equivalent, sketching of spoor evidence, enhancement, recovery and preservation of spoor evidence, case note preparation and report writing, legal considerations and courtroom testimony.

Endorsements

All applicants for certification must submit two letters of endorsement.

1. If the applicant is employed by a public law enforcement agency, one letter shall be from a superior within the applicant's department or agency and one shall be from a member of the Criminal Justice System who has used that person's services.
2. If the applicant is in the private practice, one letter shall be from a former supervisor or professional colleague and one shall be from a member of the Criminal Justice System who has used that person's services.

Tracker Certification Process

Complete the application, attach fees and mail to the address on the application

1. Application is reviewed by the Certification Board for approval.
2. Applicant will receive notification of the Board's approval or disapproval of the application.
3. Testing date and location will be provided to the applicant.

A written test must be taken. (70% is passing)

1. The written test consists of 50 multiple choice questions including but not limited to terminology, OCISE algorithm or equivalent, sketching of spoor evidence, enhancement, recovery and preservation of spoor evidence.
2. One hour is allotted to complete the test.
3. A candidate can prepare for the written test by reading: Recommended Focus Readings for Tracker Certification.
4. The test will be offered at all annual association conferences.
5. The test may be offered at other sites.

Upon successful completion of the written test, the applicant will proceed to the practical examination (tracker field problems) (passing percentage from 60 to 100 will determine level of certification).

Experience Requirements

When a tracker has developed the required level of expertise, they will be awarded a certification level. There are six levels for recognition of technical skill:[1]

LEVEL ONE

Certified Apprentice Tracker I (CAT1)

LEVEL TWO

Certified Apprentice Tracker II (CAT2)

LEVEL THREE

Certified Apprentice Tracker III (CAT3)

LEVEL FOUR

Certified Journeyman Tracker (CJT)

LEVEL FIVE

Certified Senior Tracker (CST)

---

[1]This recommended tracker certification standards was developed from review of the U.S. Marshal District of Alaska Tracking Unit tracker certification, Joel Hardin Professional Tracking Services tracker certification, and CyberTracker tracker evaluations at http://www.cybertracker.org/index.php?option=com_content& view=article&id=65&Itemid=46

LEVEL SIX

Certified Master Tracker (CMT)

In addition, there are five categories of Examiners:

Certified Examiner I (CE1)

Certified Examiner II (CE2)

Certified Examiner III (CE3)

Certified Examiner IV (CE4)

External Examiner (EE)

EVALUATION PROCEDURES

The tracker evaluation procedure consists of two distinct parts. The first is correct tracking applications and the second is correct spoor analysis.

Tracking Applications

The tracking applications evaluation is done in reasonable terrain following a human quarry (CAT) or a human quarry over more difficult terrain for the advanced certifications (CJT, CST) depending on specialty.

There are five aspects of examination for the applications of tracking:

Spoor Management- The ability of the tracker to recognize and interpret spoor. Indicators may include:

- Spoor confirmation is accurate and at appropriate times.
- Appropriate analysis and interpretation of quarry behavior from spoor.
- Can recognize and interpret both ground and aerial spoor even over hard substrate.
- Recognizes when there is no spoor and why.
- Notices spoor contamination and interpretations are accurate; manages crime scene effectively.

Trail Management- The ability of the tracker to anticipate quarry actions and therefore the quarry spoor. Indicators may include:

- Uses the Tracker's Triangle for trail maintenance.
- Uses the Extended Spoor Area to follow the trail.
- Moves the trail at a smooth pace-not to fast, not to slow.
- Institutes appropriate Lost Trail Procedures.
- Ability to re-establish trail after losing it.

Risk Management- The ability of the tracker to mitigate and anticipate trailing dangers. Indicators may include:

- Demonstrates knowledge of quarry behavior (security halt procedures, etc.).
- Recognizes changes in quarry behavior indicating danger.
- Monitors the wind for minimizing and accentuating scent.
- Leaves the trail using the offset method when danger is present and mitigates all danger to himself and team.
- Can determine the location of the quarry and other dangerous animals without risk to self and team.

Terrain Management- The ability of the tracker to understand OCOKA and project the trail to terrain for mission enhancement. Indicators may include:

- Uses terrain to project quarry movements.
- Uses cover and concealment to approach quarry and withdraw from quarry, if AT.
- Uses listening and security halts as needed.
- Notices and minimizes alarm and warning calls of local animals.
- Employs additional scouts and trackers to move the trail along.

Security Management- The ability of the tracker to take appropriate security measures for the trailing mission. Indicators may include:

- Appropriately negotiates the signs of environmental and quarry dangers.
- Institutes noise and light discipline (hand/arm signals, etc).
- Maintains appropriate Time Distance Interval for mission.
- Quarry remains unaware of scout or tracker.
- Tracker sees quarry before quarry notices tracker.
- Does not place tracker, team, or examiner in danger.

In each of these aspects, the tracker will be given points from 0 to 4:

Poor-0/1 point

Fair-2 points

Good-3 points

Superior-4 points

Since the maximum number of points would be 100, the total number of points scored would be expressed as a percentage for the application of tracking. Depending on circumstances, some indicators may not apply to the field examination. For example, if only 19 of the 25 indicators were scored, the total score would be multiplied by 25/19 to obtain a percentage.

Note: To obtain the CAT-I, the tracker must not lose more than 40 points to get 60%. To obtain the CAT-II, the tracker must not lose more than 35 points to get 65%. To obtain the CAT-III, the tracker must not lose more than 30 points to get 70%. To obtain the CJT, the tracker must not lose more than 20 points to get 80%. To obtain the CST, the tracker must not lose more than 10 points to get 90%.

Safety: If the tracker violates security and/or risk management no matter how well they do on spoor, trail, and terrain management, thus putting the tracker, team, or examiners in danger, the tracker will automatically fail and not be awarded any level of tracker advancement.

SPOOR ANALYSIS

In the spoor analysis evaluation, the candidate is awarded points as follows:

EASY SPOOR

One plus point for correct analysis (or)

Three minus points for a mistake

DIFFICULT SPOOR

Two plus points for correct analysis (or)

Two minus points for a mistake

EXTREMELY DIFFICULT SPOOR

Three plus points for correct analysis (or)

One minus points for a mistake

The total number of correct plus points are divided by the sum of all the correct plus and wrong minus points and expressed as a percentage. Not more than 20% of spoor tested may be easy spoor and not more than 20% may be very difficult spoor. The tracker may not be negatively scored based on spoor that is unreasonable.

SPOOR GUIDELINES

Easy is spoor that is a mixture of class one, class two, and some class three prints of man. These classifications are unmistakeable.

Difficult is spoor that is class four prints of man. This spoor is not specific enough due to the surface soil and substrate (soft sand or hard rock). Difficult spoor requires a trackers ability to interpret spoor formation in difficult substrate making identification difficult.

Extremely difficult is spoor that is class five prints of man. In other words, just sign with no distinction to man or animal. This spoor requires extensive experience to identify.

NOTE:

      Spoor analysis can be complex given the unpredictable nature of terrain, substrate, situation, and foot registration.

      A spoor is classified as unreasonable if two or more trackers cannot agree on the proper analysis.

      The External Examiner (EE or Echo Echo) may use their discretion to disallow any question that may be classified as unreasonable. If a dispute develops between the examiner and the tracker being evaluated or the examiner cannot explain and clearly justify the spoor to the tracker then the question may be quashed.

MINIMUM REQUIREMENTS FOR APPRENTICE TRACKER, JOURNEYMAN TRACKER, SENIOR TRACKER, MASTER TRACKER, AND EXAMINER

Certified Apprentice Tracker I (CAT1)

The CAT I Examination must be conducted by a CJTE II or above or by a CST or CMT who is familiar with the evaluation standards.

SPOOR ANALYSIS

Must be able to interpret the spoor of man and must have a fair knowledge of quarry behavior.

TRACKING APPLICATION

Must be fair in tracking applications and must have a fair ability to apply the Trail Erosion Computation (TEC).

QUALIFICATIONS

To qualify for Level I the tracker must obtain 60% for the spoor analysis portion for at least 25 signs and at least 60% for the tracking applications portion.

Must have attended a basic tracking course and satisfactorily pass the tracking written exam with a 60%.

Certified Apprentice Tracker II (CAT2)

The CAT II Examination must be conducted by a CJTE II or above or by a CST or CMT who is familiar with the evaluation standards.

SPOOR ANALYSIS

Must be able to interpret the spoor of man and must have a fair knowledge of quarry behavior.

TRACKING APPLICATION

Must be fair in tracking applications and must have a fair ability to apply the Trail Erosion Computation (TEC).

QUALIFICATIONS

To qualify for Level II the scout must obtain 65% for the spoor analysis portion for at least 25 signs and at least 65% for the tracking applications portion.

Must have attended a basic tracking course and satisfactorily pass the tracking written exam with a 65%.

Certified Apprentice Tracker III (CAT3)

The CAT III Examination must be conducted by a CJTE II or above or by a CST or CMT who is familiar with the evaluation standards.

SPOOR ANALYSIS

Must be able to interpret the spoor of man and must have a fair knowledge of quarry behaviour.

TRACKING APPLICATION

Must be fair in tracking applications and must have a fair ability to apply the Trail Erosion Computation (TEC).

QUALIFICATIONS

To qualify for Level III the tracker must obtain 70% for the spoor analysis portion for at least 25 signs and at least 70% for the tracking applications portion.

Must have attended a basic tracking course and satisfactorily pass the tracking written exam with a 70%.

Certified Journeyman Tracker (CJT)

The CJT Examination must be conducted by a CJTE III or above or by a CST or CMT who is familiar with the evaluation standards.

SPOOR ANALYSIS

Must be able to interpret the spoor of man and must have a fair knowledge of quarry behavior.

Must be able to interpret the spoor of man and must have a fair knowledge of quarry behavior.

TRACKING APPLICATION

Must be good in tracking applications and must have a good ability to apply the Trail Erosion Computation (TEC).

Must be able to track man and be able to overcome common soft counter tracking methods.

QUALIFICATIONS

To qualify for Level IV the tracker must obtain 80% for the spoor analysis portion for at least 35 signs and at least 80% for the tracking applications portion.

A minimum of three years' experience in tracking application is required.

Certified Senior Tracker (CST)

The CST Examination must be conducted by at least one Examiner who is familiar with the local terrain and an External Examiner who is familiar with evaluation standards in other areas. At least one should be a Master Tracker.

SPOOR ANALYSIS

Must be able to analyse the spoor of all quarry.

Must have a very good knowledge of quarry behavior.

The spoor analysis portion for the CST should not only be more rigorous, but also put greater emphasis on projection and interpretation of human behavior from spoor.

TRACKING APPLICATION

Must be superior in tracking applications and have a strong ability to project the trail.

This projection of the trail includes the ability to predict where spoor will be found beyond the immediate area.

Must have a superior ability to apply the Trail Erosion Computation (TEC).

Must be able to track man and be able to overcome most soft counter tracking methods and common hard counter tracking methods

Must be able to detect signs of stress from spoor.

QUALIFICATIONS

To qualify for Level V the tracker must obtain 90% for the spoor analysis portion for at least 50 extremely difficult signs and at least 90% for the tracking applications portion.

The CST must also pass an oral test on human behavior which will focus on depth of knowledge as well as the written exam on the technical aspects of spoor analysis and tracking applications.

A minimum of six years' experience in tracking application is required.

The CST will be qualified to train new trackers alone as well as specializing in particular aspects of the art and science of tracking.

Certified Master Tracker (CMT)

The CMT must have an excellent knowledge of human behavior and be capable of a highly refined interpretation of spoor in difficult terrain.

Must have originality and creative insight and must have well-developed intuitive abilities. His or her expertise would be equivalent to that of the best tradesman of any ancient or modern craft.

The CMT certificate will be awarded to senior trackers with at least twelve years' experience in tracking application and who have demonstrated an ability to make pivotal aid to our understanding of tracking, crime scene, and quarry behavior.

The CMT will be qualified to train and evaluate Trackers and Senior Trackers.

The CMT will involve a process of peer review, involving two Certified Master Trackers familiar with the local terrain and an External Evaluator who is familiar with evaluation standards in other areas.

Examiners

Certified Examiner I (CE1)

QUALIFICATIONS

To qualify for the Certified Examiner I certificate the candidate must obtain 70% for the Spoor Analysis evaluation and at least 70% for the Tracking Applications of a human trail that is not easy to follow.

EXPERIENCE

A minimum of one year experience in tracking application is required and they must be familiar with the evaluation standards.

NOTES

The CE1 is not qualified to conduct a CJT Evaluation and can only assist a CST or CMT to conduct an evaluation.

The CE1 examination must be conducted by a CE2, 3, or 4 who is familiar with the relevant examination standards.

149

Certified Examiner II (CE2)

QUALIFICATIONS

To qualify for the Certified Examiner II certificate the candidate must obtain 80% for the Spoor Analysis evaluation and at least 80% for the Tracking Applications of a human trail that is not easy to follow.

EXPERIENCE

A minimum of three years' experience in tracking application is required and they must be familiar with the evaluation standards.

NOTES

The CE2 is not qualified to conduct a CJT Evaluation and can only assist a CST or CMT to conduct an evaluation.

The CE2 examination must be conducted by a CE3, or 4 who is familiar with the relevant examination standards.

Certified Examiner III (CE3)

QUALIFICATIONS

To qualify for the Certified Examiner III certificate the candidate must obtain 90% for the Spoor Analysis evaluation and at least 90% for the Tracking Applications of a human trail that is not easy to follow.

EXPERIENCE

A minimum of five years' experience in tracking applications is required and they must be familiar with the evaluation standards.

NOTES

The CE3 is not qualified to conduct a CST Evaluation and can only assist a CMT to conduct an evaluation.

The CE3 examination must be conducted by a CE4 and/or an EE who are familiar with the relevant examination standards.

Certified Examiner IV (CE4)

QUALIFICATIONS

The examiner must be a CST or CMT and must have mastered the method of evaluation described in this document.

EXPERIENCE

A minimum of ten years' experience in tracking applications is required and they must be familiar with the evaluation standards.

NOTES

In addition the CE4 must have a clear understanding of the standards required in distributing points on spoor analysis and the various areas of tracking applications and must have an in-depth knowledge of the local terrain.

External Examiner (EE)

The External Examiner must be a CST or a CMT and must be familiar with evaluation standards in other areas. The role of the External Examiner is to ensure that consistent standards are maintained in different areas and over time. This requires both an understanding of local tracking conditions and how standards would translate to different tracking conditions in other areas.

TRACKER CERTIFICATION
OFFICIAL RECORD [1]

CERTIFIED APPRENTICE TRACKER I   II   III ____   CERTIFIED JOURNEYMAN TRACKER_____
CERTIFIED SENIOR TRACKER ____   CERTIFIED MASTER TRACKER_____

ADMINISTRATIVE DATA

NAME:		SSN:	GRA DE:

TRACKING LEVEL:	ORGANIZATION:

TYPE OF REPORT	PERIOD OF CERTIFICATION			RATED MONTHS	
	FR:		TO:		

DUTY DESCRIPTION

PRIMARY DUTY TITLE:	SECONDARY DUTY TITLE:

DESCRIPTION OF TRACKING RESPONSIBILITY FOR THIS EXAMINATION:

EVALUATION OF PERFORMANCE
1 = ATTEMPTED, BUT INEFFECTIVE;   2 = EFFECTIVE;   + = DISCIPLINE, PATIENCE, & CONTROLLING

Examiner:	Endorser:	Tracking Competence:
		1. Understands and utilizes the pre-trailing checklist.
		2. Effectively negotiates the Initial Trail Assessment.
		3. Properly sends MINTREP to higher with mission essential elements.
		4. Uses the Trail Confirmation Standards to effectively back or if necessary change the Initial Trail Assessment.
		5. Logs sufficient detail for any Register Deviation Anomaly.
		6. Performs Slope Topography and Grade LOS.
		7. Performs field sketches to diagram essentials along the LOS.
		8. Keeps a detailed observation log with a chronology of events from start of mission to end.
		9. Understands and utilizes the Post-trailing checklist.
		10. Works well as a full team member.

For Official Use Only

Experience Rated PASS FAIL	Minimum Requirements Met: YES NO
Education Rated PASS FAIL	Field Exam Date:
Written Exam Date: Score:PASS FAIL	Score: PASS FAIL
	Examiner Comments:

Signature of Certified Examiner_____ Date

Signature of Certifying Official_____Date

Source: Modified form from the U.S. Marshals District of Alaska Tracking Unit.

TRACKING FIELD EXAMINATION SHEET

Date:	Quarry Type:
Tracker:	Quarry Size:
Examiner(s):	Temperature:
Location:	Precipitation:
Start Time:	Visibility:
End Time:	Cloud Cover:
Duration of Exam:	Quarry Found:

EXAMINATION CRITERIA

Spoor Management	
	Spoor confirmation is accurate and at appropriate times
	Appropriate analysis and interpretation of quarry behavior from spoor
	Can recognize and interpret both ground and aerial spoor even over hard substrate
	Recognizes when there is no spoor and why
	Notices spoor contamination and interpretations are accurate; manages crime scene effectively
Trail Management	
	Uses the Tracker's Triangle for trail maintenance
	Using the Extended Spoor Area to follow the trail
	Moves the trail at a smooth pace-not to fast, not to slow
	Institutes appropriate lost trail procedures
	Ability to re-establish trail after losing it
Risk Management	
	Demonstrates knowledge of quarry behavior (security halt procedures, etc)
	Recognizes quarry behavior indicating danger
	Monitors the wind for minimizing and accentuating scent
	Leaves the trail using the offset method when danger is present
	Can determine the location of the quarry and other dangerous animals w/o risk
Terrain Management	
	Using terrain to project quarry movements
	Uses cover and concealment to approach quarry and withdraw from quarry
	Uses listening and security halts
	Notices and minimizes alarm and warning calls of local animals
	Employs additional trackers to move the trail along
Security Management	
	Appropriately negotiates the signs of danger
	Institutes noise and light discipline (hand/arm signals, etc)
	Maintains appropriate Time Distance Interval for mission
	Quarry remains unaware of tracker
	Tracker sees quarry before quarry notices tracker

OBSERVATIONS:

153

SPOOR ANALYSIS EXAMINATION SHEET										

Examiner(s):						Date:		Location:		

Quarry	Spoor	C-1	C-2	C-3	C-4	C-5	Spoor	A	B	C
Results										

APPENDIX E

RECOMMENDED TRACKER VOCABULARY

This thesis contains many spoorology specific terms. Yet, there is some overlap with the more common tracker terms in general. This appendix is a general tracker vocabulary developed through review of all available research literature contained in the bibliography.

Accidental Damage Characteristics - Random features such as nicks, cuts and gouges made to the out sole during wear, or, deposited before molding.

Aerial Cone of Entry/Exit - The downward and upward line of foot travel leading from the ground to the apex of foot arc and the return to ground.

Aerial Spoor - Any visible disturbance to vegetation above ground level which indicates the passing of the quarry you are following. Aerial sign often needs to be verified by ground spoor. Also known as top spoor.

Aging - Process of determining time lapse since spoor was made, considering vegetation damage, rain, new snow cover, sun, other human or animal passage, and other natural elements.

Angle of vision - Act of getting lower to the ground or rising higher in an attempt to obtain a better view of spoor.

Apex of Foot Arc - The highest point of foot elevation from one step to another.

Arch - The bottom part of a shoe last from heel to width.

Associative - A determination that the area under examination shares some features with the known shoe, or, any other shoe having the same physical, size, mold, design, and pattern features.

Associated Sign - Sign a quarry leaves along its trail other than its tracks.

Asymmetrical gaits - There is an uneven spacing of footfalls and the right half of the track pattern differs from the left.

Backing - Walking backward in all attempt to camouflage the actual track by presenting an incorrect direction of travel. A Soft counter-tracking method.

Back Trailing Method - Also known as back tracking. Following the trail in reverse back to the start point.

Ball - Fleshy part of the foot just back of the toes.

155

Baseline - The basis from which you judge everything else, the undisturbed area from which you operate to find disturbances within. Nature will continue to establish and alter baselines.

Bipedal - Moving on two feet.

Bound Trailing Method - The deliberate movement of progress on the trail from one possible track trap to another within the line of sign. This method is used in conjunction to the Deliberate Trailing Method to move the trail along.

Bracketing - The act of maintaining continuity and constancy of sign by skipping the next spoor in line to a spoor in front of the last known track, then using it to find the one skipped.

Broken Twigs - Small particles of vegetation or twigs, broken in such a way as to indicate damage from human footwear.

Bruising - Damage to vegetation by someone stepping on it.

Brushing Out - Using a branch, grass, or clothing article in an attempt to brush or erase spoor from an area. A common method of soft counter tracking.

Camouflaging - The concealment of the quarry and trail by means of making it appear to be part of the natural surroundings.

Class Characteristics - Intentionally created features exhibited by similar items that will be in common, generally appearing as a result of manufacture.

Clue - A piece of evidence that may or may not contribute to the interpretation process of understanding the trail of a quarry.

Color - Refers to wavelengths of light as seen by the eye and interpreted by the brain. Valuable in differentiating one object from another. In nature, color tends to be muted and less vivid.

Color change - A difference of color, value or texture from the area that surrounds it.

Communicating - The exchange of information gleaned from the spoor with others that need to know.

Comparative Examination - A process of analysis in which an item of unknown origin designated as the 'questioned item, is compared by its features to objects designated as known standards. This is a process in which conclusions of limitation, distinction, elimination, and identification can be demonstrated and verified.

Compressed Areas - 1. Areas of ground surface compressed in a manner that indicates humans as the cause; 2. Area of soil or vegetation where the surface is compressed into/onto the apparent 'lay of the land". Giving the appearance of a depression or compaction of the ground surface by a footfall.

Conclusive Sign - Spoor that when considered on its own, can with no doubt, be linked to the quarry being tracked.

Confirmatory evidence - When following aerial spoor the tracker must constantly look for ground spoor or other sign to confirm the disturbance being followed was made by the quarry.

Consistency of Sign - The tendency of spoor evidence to remain relatively the same, or similar, throughout each step, in an area where the conditions do not change significantly.

Contamination - Other spoor from police officers, crime scene personnel, family members, other trackers, other emergency personnel, and animals that obscures or completely obliterates a quarry's spoor.

Continuity of Sign - The evidence of footfalls in proper sequence along a line of sign, generally unidentifiable.

Contrast - Refers to any difference in color, shape, texture, or composition between something and its surroundings. Examples include transfer, shine, and disturbance of ground vegetation. The greater the contrast, the more compelling the cue. Contrast is used extensively in spoor aging, by making a fresh reproduction of the disturbance and comparing the contrasts of the existing disturbance.

Converge - To approach a common point from different directions.

Corroborate Sign - Spoor that is not decisively human, and could have been caused by another animal. Not conclusive on its own, but is used with other evidence to make conclusive decisions.

Counter tracking (Hard) - The process and art of stopping or slowing trackers efforts to follow spoor.

Counter Tracking (Soft) - The act of camouflaging or otherwise hiding spoor.

Crease - Sign, usually left on soft leaves or other vegetation, which resembles a straight fold on paper.

Crying - The natural weeping of vegetation fluids resulting from footfall damage, body passage, and deliberate breaking of twigs and branches.

Cutting - Once the tracking operation is underway, an attempt to intercept the trail is re-employed to speedily advance the spoor great distances; An operation used principally along natural barriers to locate spoor; Method used to locate a lost trail by moving ahead in the direction of travel and attempting to relocate the track by walking diagonally to the previous route.

Cut Trailing Method - The method of finding the trail of a quarry by traveling perpendicular to the suspected travel route of the quarry.

Decay Pit - An area of close examination of track deterioration in a controlled environment of study. This area is used by a tracker to watch the aging of spoor on various surface soils and vegetation.

Deception - The attempt to confuse, disguise, or conceal spoor by ; walking backwards (backing), brushing out, laying false trails, or other means, to deceive or confuse direction of travel, number of persons, or presence of spoor.

Degree of Wear - The extent that a particular portion of the shoe has been worn.

Deliberate Trailing Method - The slow, methodical precise accumulation of all trail evidence, overlooking nothing on the trail to establish linearity.

Diagnostic - Identification of spoor, characteristics, etc.

Direction of Travel - The navigational direction of the forward movement of the trail. Usually a North, South, East, West or by the degree of compass heading.

Discardable - Littering, primarily of the "garbage" variety.

Displacing - The change from one place to another.

Distortion - An unclear or inaccurate representation of the shoe outsole in the depression due to interference with the impression making process. There are landscape, mechanical, self-imposed, and weather.

Disturbance - Any evidence of recent change or rearrangement

Diverge - To move or extend in different directions from a common point.

Doorway - A natural construction in nature, such as trees or rocks, which presents an opening to walk through similar to a door. One of the ten parts of the tracker's triangle.

Dynamic Trailing Method (DYTM) - The explosive trailing method that decreases the Time Distance Interval between the tracker and quarry. Also known as tactical tracking or combat tracking.

Elimination Method - This method is application of the process of working outside in. The trail can be maintained by eliminating where the quarry did not go. This method is used when the trail is over hard to read substrate.

Endemic - Native or limited to a particular location.

Enhancement - Chemical, photographic or physical means used to render an impression visible or more distinctive in appearance.

Erroneous Association - The incorrect determination that the unknown impression and the known shoe sufficiently share features in the category(s) of pattern, size, design, and/or wear features.

Erroneous Identification - The incorrect determination that the unknown impression was made by a known shoe.

Evidence - Something legally acceptable before a court; Facts or knowledge that will prove or disprove an allegation; An object or witness that bears on or establishes an issue.

Extended Spoor Area - The area in front of the Primary Spoor area or the farthest most point in the tracker's triangle where the tracker is looking to trail the quarry. One of the ten parts of the tracker's triangle.

False Trails - Leaving a good trail or spoor into a poor spoor area, then departing on another route. Method of soft counter tracking.

Flagged - Leaves or grass turned in the direction of travel, showing their underside surfaces.

Flat - Footgear with no discernible heel, i.e., the shoe tread is flat across its entire length.

Flattening - The leveling of soil, pebbles, twigs, rocks or other substances into the ground which alters the texture of the surface producing a flattened area, which may create a shine by reflecting more light than the surrounding area. Caused by the flat sole of a shoe compacting the ground under a quarry's weight.

Flex Point - The point within the foot roll application where the foot flexes at its maximum and/or minimum points and transfers this to the foot registration in the soil and vegetation.

Focus lock - Getting "caught up" in one aspect of tracking and not paying attention to the surroundings. You should always; look twice, vary your vision by looking from different angles and alternate from looking at minute detail in and around the track to reviewing the overall situation, your location, direction of travel and look for your quarry.

Foot Axis - Imaginary line down the center of the foot.

Foot Impression Reference Matrix - Systematic calculations of quarry height, weight, foot size measurable from the ground.

Foot Measurement Analysis - The cross point sectioning of the bottom of the foot to the ground impression; thus, puzzling the foot together to form a match to a known print.

Footprint - Visible evidences that a foot has contacted a surface.

Footprint Classification or Classification of Spoor - The foot analysis by category to establish expected quarry.

Foot Roll - The application of foot movement as weight is placed during motion. The heel to toe motion and side-to-side motion for stabilization and progression. This contains the impact point, flex point, pivot point, and terminal point of foot motion.

Fore - front

Fore foot - The leading foot in a two-step set.

Gait - The pace that is traveled during human locomotion; The position of the footprints in relation to each other; Each gait leaves a coordinated pattern of movements.

Gateway - A natural construction in nature, such as trees or rocks, which presents an opening to walk through similar to a gate. One of the ten parts of the tracker's triangle.

Gathering (Information) - The complete ongoing process of collection of spoor information.

General Class Characteristics - Basic design features in an outsole pattern that are totally indistinguishable between different outsoles.

General Condition - The overall state or general amount of wear on a shoe.

General Course of Action - The overall "what must get done" during tracking. This is in terms of the tracker, team, as well as the quarry.

Grass Trail - The bending, and intertwining of grass or brush indicating quarry passage.

Ground Contact Points - (1) The visual points within and around the print which tells how the quarry foot contacted the surface soil and vegetation. (2) Physical disturbances found in and around a track, which indicate action such as pressure, speed, twisting etc. There are indicator pressure releases and fluctuating pressure releases.

Ground Hardness Type - The medium of surface soil giving the tracker a mental picture of how the foot would interact with gravity and age.

Ground Spoor - Any imprints, mark, indentations, transfers or vegetative damage found on the ground which can be positively identified as a disturbance left by the quarry you are tracking. Also known as bottom spoor.

Healing - The process with live vegetation in which damage is naturally repaired. Used to age sign.

Heel - Footgear with a clearly discernible heel, such as a cowboy boot or low-quarter dress shoe.

Heel Marks - The curved mark or depression on the ground surface made by the walking motion of the heel part of human footwear.

Hind - back

Hind Foot - The trailing foot in a two-step sequence.

Identification - The determination that an area under examination on the unknown impression was made by the known shoe to the exclusion of all other shoes.

Impact Point - The precise point the foot makes contact with the ground and usually reveals the most distinctive information.

Impact Point Gradient - The angle of foot movement from the apex downward to contact with the ground.

Impression - A three-dimensional print in some soft material, that displays the size, design and wear features present on the outsole of an item of footwear.

Indexing - The comparison of the know spoor(represented either by foot or thumb indexing) and comparing this to the surrounding area and the questioned spoor.

Individual Characteristics - Those characteristics which are unique to a given object and set it apart from similar objects; Randomly created features that are individualistic, peculiar and/or distinguishing to a specific outsole. These appear as a result of wear and use of footwear.

Initial Commencement Point - The place where the actual tracking operation begins.

Initial Trail Assessment - The cursory collection of spoor data to establish the generals of the quarry trail.

161

Intelligence Preparation of the Trail - The complete ongoing collection of trailing information that prepares the tracker for the trailing mission.

Intermittent Attention - A constant refocusing between minute details of the track (Micro) and the whole pattern of the environment (Macro).

Inventing Spoor - A condition, usually caused by fatigue, where the tracker "sees" spoor that isn't there. To fabricate spoor in the mind's eye (operating from preconception, not perception).

Irregularity - A disturbance that disrupts baselines or templates.

Jump Trailing - A form of tracking that involves finding an obvious spoor, then proceeding along the assumed direction of travel until another obvious spoor is found. A highly risky endeavor that may result in contamination of the area and ultimate destruction and loss of the line of sign. Also known as jump tracking.

Known Footwear - The shoe(s) of known origin that is/are compared to a questioned footwear impression.

Last Known Point - The most recent location the quarry can conclusively be said to have been, based on all available evidence, including human sign, eyewitnesses, vehicles or personal gear, sign-in or summit logs, etc. Differs from Point Last Seen (PLS) in that a PLS requires the seeing of the quarry by another human.

Latent footwear impression - A questioned or unknown two dimensional footwear impression that is to be compared to the known footwear impression.

Left Right Analysis - The system of calculating the foot placement in soil and reference it to said placement to trail particulars.

Length of the Track - From heel to end of toe.

Lift - The process of transferring an impression from its original surface to a transportable surface, for the purpose of its preparation for comparative examination.

Lighting - Use the correct angle of the primary light source in order to ensure optimum visual inspection capabilities. Light can be altered and improvised. Objects will absorb and reflect light in differing degrees.

Limited Class Characteristics - Manufacture or design features that can separate an object from others in its class. Limiting features or variations that can reduce the field of possible known candidates.

Limited Value for Comparison - Some aspect of the impression, or, the recovery process has limited the extent to which the impression may be compared to the known standard.

Limp - When one foot consistently takes a shorter stride than the other.

Line of Sign - The continuity and constancy of spoor evidencing the passage of a quarry.

Litter - Any unnatural debris, garbage, cigarette butts, paper, etc., which may indicate human passage. (Litter may or may not be related to the quarry being tracked).

Location, Habitat, Season - Three factors that must be considered when estimating quarry and tracker interaction with the environment.

Lost Trail Procedures - A systematic approach to finding the spoor when it is no longer immediately available. Also known as lost spoor procedures.

Mark(ing) - Highlighting the presence of a line of sign utilizing a standard system of scuff marks, engineer tape, colored strips on wire, or wooden pickets, so you or others can find it again.

Mechanical Stride Zero - The experienced average of stride of all tracking operations for a particular quarry.

Milling - A static position where the quarry moves about in place.

Milling Pad - The worn down piece of substrate from the process of milling.

Natural Barriers - Areas such as streams, banks, roads, railroad grades, etc., which generally interrupt a quarry☐s passage and show sign well.

Natural Lines of Drift - The lay of the land that influences human and animal movement patterns. One of the ten parts of the tracker's triangle.

Non-visible sign - Information you can acquire by smelling, hearing and touching.

Normal Walk Gait - The common movement application of locomotor movement.

On-Trail - The physical act of a tracker, following a set of tracks on the ground made by a quarry. Also known as a Follow-up.

Outline - The edge of an object, it marks the boundary or perimeter line of a shape. Outline items may be a small line or a complete track. Examples include heel curves, heel shape, or toe shape.

Pace - 1. Measurement from a point in a given footstep to the next time that same point occurs, e.g., from the tip of the toe on the right foot to the next tip of the toe on the right foot; 2. A single step measured from the heel of one foot to the heel of the next foot.

Path of Least Resistance - The routes within nature that are easiest physically and instinctively to follow. One of the ten parts of the tracker's triangle.

Pattern - The spatial arrangements of the spoor.

Pigeon Toed - A quarry that walks with their toes pointed inwards.

Place Last Seen - Place where a witness through direct visual observation; The most recent location where the quarry was known to have been seen by another human.

Perimeters Cut - A cut trailing method of limiting a search area or locating spoor along natural barriers. Usually a natural boundary is used perpendicular as possible to the line of travel.

Phalanges - Toe bones of the foot.

Physical Environmental Restrictions - The natural terrain topography and weather that influences the tracker and quarry.

Pitch - The distance the front of the print deviates inwards or outwards from the line of travel. Also known as Angle of Track.

Pivot Point - The point within the foot roll application where the foot pivots at its maximum and/or minimum points for direction change and transfers this to the foot registration in the soil and vegetation.

Primary Spoor Area - The area of correct size and location in relationship to other spoor, in which the next spoor should be located. One of the ten parts of the tracker's triangle.

Print - An imprint left on the ground by a single foot.

Pressure release - The physical deformities found in and around a track which were created by the energy of the foot as it contacted the ground and the pressure it exerted as it left the ground. There are two: indicator and fluctuating.

Quadrant Reference Analysis - The ten quadrants of foot mapping that enable the tracker to piece the known foot together while on the trail.

Quarry - 1. The person, animal or object being pursued; 2. An alternative to enemy, fugitive, subject, party, target or "the pursued."

Register Deviation Anomaly - This is the analysis of all deviations from the normal movement of the quarry. Must be analyzed from macro to micro.

Regularity - An effect caused by straight lines, circles, or other geometric shapes pressed into the ground leaving marks that are not normally found in nature.

Rhythm - Movement or variation characterized by the regular occurrence or alternation of different qualities or conditions. A regular or harmonious pattern created by lines, shapes, colors, values and textures.

Rotatory - Circle

Route Influencers - Things that will induce a quarry to walk a certain route.

Running - Generic for expedient movement. It contains longer strides and is implemented by the acceleration of locomotion.

Sand Trap - Dirt areas, occurring naturally or manmade which by their own nature show sign well. Also known as track trap.

Scatology - The science of studying scat.

Scuff Mark - The mark or sign caused by footgear contacting the ground surface. Can also be the transference of sole material onto a surface.

Secondary Spoor Area - The area next to the Primary Spoor Area on both sides within the tracker's triangle.

Shadow - An area that is not or partially illuminated due to blockage of the light source.

Shape - 1. Refers to anything unusual to a given environment that corroborates spoor. Examples include (larger than a nickel.) areas of flattening, and curves that don't correspond to animal tracks (such as a heel curve or edge of a sole); 2. The form of an object, important element for recognizing a track (compressions).

Shine - Light reflecting off the ground surface where foot compression has created a flatter surface. That area will reflect more light than the surrounding area. Shine is unique in that it is one item of spoor that is often best seen from a distance, and is invisible when viewed close up.

Sign - Evidence of a person or animals passage; Any evidence of change from the natural state that is inflicted on an area by the passing of a person, animal or object; Non-conclusive evidence of passage or movement of a person, animal or object; Other indicators of a quarry's passage that does not fall into the categories of ground or aerial spoor.

Signature Track - A single footprint or piece of evidence that clearly displays unique characteristics so as to be unmistakably identifiable.

Silhouette - The outline of an object that appears dark against a light background.

Slack - The security observer designated to provide close protection for the tracker.

Sole - The whole bottom of the foot, to include hair, no hair, and pads.

Spacing - The result of arranging by intervals.

Splay Footed - Subjects who walk with their toes pointed outwards.

Spoor - 1. A set of tracks laid upon the ground and visible to the tracker. Spoor is totally interchangeable with the words "tracks," "trail," or "set of prints." 2. The collective amount of evidence left by human passage, including tracks, scent, litter, body fluids and substances, and human witnesses; The track or trail of a wild animal.

Spoor Conscious - To understand and appreciate that when any human moves on the surface substrate they leave evidence of their passage.

Spoor Interpretation - To understand and comprehend the meaning of what the tracker sees on the ground. To understand the story written on the ground. The interaction between the tracker and the trail.

Spoorology - The science of studying spoor evidence.

Spoor Pit - An area of close examination of track and pattern in a controlled environment of study. This area is used by the tracker to tune the eyes up and the create controlled stories of animal and human movement.

Staining - The discoloring or tainting of a substance. Like water of a rock or mud on leaves, etc.

Stalking - The art of moving up on a quarry without the quarry being aware of your approach.

Stick Measuring Device - Known as "smiddy". This tool is valuable in keeping measurements of the trail. Also known as a tracking stick. This tool is only used on a non-tactical or non-dangerous trail of the quarry such as an in-active crime scene. See Tracking Quirt.

Straddle - Measurement taken of the undisturbed area between the opposite feet at the Innermost points could be a negative or positive measurement. (Decreases with speed.) The distance between a quarry's feet on either side; Distance between opposite feet, measured from the center of each foot; Measurement taken perpendicular to the line of travel at the widest point of a trail or group pattern. Includes width of the tracks (trail width).

Straddle, Pitch, Width - Three factors that assist in understanding quarry movement on the ground.

Step - Measurement from the tip of the toe to the back of the heel on the next step. Stride increases with speed.

Stride - The measurement taken from the heel of one foot impression to the heel of the next foot impression on people. Toe to toe on animals. (Heels tend to be the most visible disturbance found when tracking people. Toes and hoof points on animals are more prominent.) The distance one foot travels in a single step, measured from the same point on each print.

Stride (other) - Measurement between individual tracks. Measured Heel to heel.

Subject - Another name for the quarry being followed or tracked.

Substantiating Sign - Spoor which is insufficient in itself, without other confirming factors, to prove it was made by the quarry being tracked.

Suddenness - Movement that is quick, abrupt, and hasty.

Sun Light Angle - The angle of the sun in relation to the trail.

Symmetrical - There is an even spacing of footfalls and the right half of the track pattern differs from the left.

Symmetrical gate - The interval between footfalls is evenly spaced, and spoor patterns are symmetrical for the right and left sides of the trail.

Tactile - Using the sense of touch to aid in assessing information, usually the age of spoor.

Target Quarry Reference - The suspected quarry, either man or animal.

Templates - Various shapes colors and textures stored in the mind's eye in familiar arrangements from which you compare attractions noticed by the physical eye for identification.

Terminal Point - The point where the foot leaves the ground in preparation for the next step.

Terminal Point Gradient - The last point the foot has made contact with the ground on its upward climb to the step apex.

Texture - Refers to the consistency or smoothness of a surface, the relative roughness or smoothness of an object in relation to its surrounding.

Three dimensional impressions - Where an object presses into something soft which retains the impression of that object.

<u>Three Sixty.(360)</u> - When involved in lost trail procedures, trackers move around in a circle or box in an attempt to locate the spoor.

<u>Time Shadow Effect</u> - The time of day in relation to the cast of shadow on the foot impression.

<u>Toe Dig</u> - The indented mark or sign left in a single print from a normal walking motion when the foot propels the body forward.

<u>Toe Drag</u> - When the toe drags soil up onto the horizon of the track.

<u>Track</u> - 1. (n) An impression left from the passage of a person or animal; 2. (*V*) To follow a subject by locating and proceeding along its line of sign; 3. A mark or succession of marks left by something that has passed; 4. Awareness of something occurring or passing; 5. Conclusive evidence of a quarry's passage; 6. The imprint left on the ground from a single foot. Also known as spoor.

<u>Tracker</u> - 1. The member of a tracking team who is physically looking for and following spoor; 2. A person with the ability to follow, read, and interpret the "sign story" at a crime scene.

<u>Tracker's Triangle</u> - The first step in LTP's. The tracker uses the tracker's triangle to superimpose the mental image upon the trail which allows the tracker a mental formula to maintain the trail at all times.

<u>Tracking</u> - 1. Following a line of sign; 2. Following someone or something by stringing together a continuous chain of their spoor; 3. Finding, identifying, interpreting and following spoor of man, animal or some other quarry; 4. The science of following a line of spoor and sign that was left sometime in the past.

<u>Track Erosion Computation</u> - Also known as aging. The evaluation of time to spoor or disturbance placement in soil or vegetation.

<u>Track Pattern</u> - A distinctive arrangement of spoor.

<u>Track Trap</u> - 1. A dirt area, either naturally occurring or constructed by the tracker, composed of soft earth that will readily display spoor and signal human passage; Places that are outstanding to check for spoor because of the ease of which tracks can be seen.

<u>Tracking Team</u> - 1. A specified number of trackers, each which specific functions, who follows a line of sign; 2. A tracking team consists of a tracker, three scout observers, a security observer , and a team leader.

<u>Tracking Quirt</u> - A piece of thin rope tied to the wrist of the non-shooting hand for tactical operators. This tool is used for keeping measurements of the trail without hindering the quick application and employment of a firearm.

Trail - 1. A series of footprints, tracks, or spoor; 2. A line of prints, sign, or spoor left by man or animal.

Trailing - 1. The following of a trail; 2. Following the general route of the quarry without proceeding along the actual line of sign.

Trail Confirmation Standard - The systematic way to verify the initial trail assessment and alter changes as necessary.

Trails Leading Edge - The edge of the trail next to the tracker. This is the Last Known Track.

Trail width - Total width of the entire spoor group, measured from the outer most edge of the left foot to the outermost edge of the right foot. Also known as trough.

Transfer - The evidence of a substance - dirt, mud, debris - being carried by feet from one surface and being redeposited on succeeding footfalls. Also known as staining.

Transverse - Cross

Transitions - Spoor patterns or gaits that are altered due to; a change of speed, gait, direction or uneven surfaces.

Two Dimensional Impression - Where an object transfers an image to a surface.

Uniform Scale Format - The precise and detailed collection of trail evidence.

Value - Contrast, relative lightness or darkness, if an object is lighter or darker than its surrounding then it is more likely to be discovered.

Variable Trail Patterns - The different and varying ways man and animal can move, thus establishing patterns that can be calculated.

Varied vision - To look up often, look for patterns, alternate from looking for tracks, to scanning ahead, looking to the sides and behind and establishing the TLE. Going from parts to whole, from large to small.

Visual Search Patterns - Methods of systematic seeing to enhance visual capture of ground and aerial spoor.

Walk - To move each foot independently of the other. Indicated by two parallel rows of alternating evenly spaced prints. The slowest movement application in humans. This may not equate to most common, however. The walk consists of slow, normal, and fast.

Weathering - The mechanical chemical process of track and sign returning to its natural place. The effects of weather on quarry spoor over time.

<u>Width of the Track</u> - the greatest distance from right side to left of the foot, either over the toes or sole.

CURRICULUM VITAE

TYRON J. CUNNINGHAM, CMST

ACADEMIC EDUCATION

ASHFORD UNIVERSITY Clinton, Iowa	B.A. Social and Criminal Justice	06/2009
Alpha Sigma Lambda National Honors Society	Inducted Member for Academic Excellence 3.84 GPA	04/2009
CAPSTONE PROJECT:	The Essential Uses of Forensic Tracking in Police Investigations	

LAW ENFORCEMENT EDUCATION

22 May '06 – 26 May '06 Managing for Success # 601 U.S. Marshals Training Academy St. Simons Island, GA	Learned the advance applications of situational leadership to effectively manage and motivate investigators to foster partnerships. I learned and applied open communications skills to increase quality and frequency of conversation about performance and development of the investigator.
24 May '04 – 28 May '04 Law Enforcement Executive Management Course FBI Intermountain Law Enforcement Executive Command College West Yellowstone, MT	Learned to prepare for the challenges that I might confront in the future as a law enforcement manager. I learned the executive management model, what good leadership is to the executive, and what the social, political, and demographic trends that will affect my leadership. I learned crisis and stress management, media relations, and what executive challenges I will face in application of the law.
05 May '03 – 09 May '03 Intro. to Management and Leadership #303 U.S. Marshals Training Academy St. Simons Island, GA	Learned the basics of situational leadership and the corresponding developmental levels. I was taught to balance my role to delegate, support, coach, and direct my investigators to the competence and commitment levels of my investigators.

08 May '95 – 18 May '95
Advanced Deputy U.S. Marshal #506
Federal Law Enforcement Training Center
Brunswick, Glynn Co., GA

Learned advanced practices of execution of federal arrest warrants, parole warrants, custodial and extradition warrants. The security model for protection and security of federal judiciary, jurists, court officers, and other threatened persons in the interest of justice where criminal intimidation impedes the functioning of the federal judicial process.

25 May '91 – 03 Jul '91
Basic Deputy U.S. Marshal #104
Federal Law Enforcement Training Center
Brunswick, Glynn Co., GA

Learned the basics to enforcement of federal laws and how to support all elements of the federal justice system, i.e. executing federal court orders, investigation and apprehension of criminals, providing security for federal judges and other court personnel, assuring the safety of endangered government witnesses and their families.

25 Mar '91 – 24 May '91
Criminal Investigator Course #CI-114
Federal Law Enforcement Training Center
Brunswick, Glynn Co., GA

Learned all basic criminal investigative methods and procedures necessary for investigative competence, i.e. legal training, interviewing, search warrants, working with informants, report writing, surveillance, crime scene investigation, undercover operations, criminal intelligence, etc.

MILITARY EDUCATION

05 Apr '89 – 20 Apr '89
Light Leaders Course
AK ARNG
Ft. Richardson, AK

Trained and evaluated on my leadership ability to lead soldiers in a tactical environment according to the task, conditions, and standards of over fifty common infantry tasks. I implemented proper troop-leading procedures and accomplished my designated patrol missions. I became a seasoned leader of small-unit tactics.

15 Apr '87 – 22 May '87
BNCOC
NCO Academy
Ft. Richardson, AK

Learned leadership skills, NCO duties, responsibilities and authority, and on how to conduct performance-oriented training. I learned to be a battle competent NCO who is qualified to lead, evaluate; and counsel soldiers. I was evaluated on conducting soldier individual and collective training; and to be a teacher of leadership values, attributes, skills, and actions. I learned leadership skills from instructors who assessed my leadership potential and evaluate my ability to apply lessons learned to effectively lead soldiers. I was provided an opportunity for education, to learn warfighting skills, and to gain experience.

16 Jul '86 – 15 Aug '86	Learned to be focused on NCO core areas of leading, training, maintaining standards, caring for soldiers, technical competencies and tactical warrior skills. I learned to be a small unit leader who can employ all warfighting functions rapidly. I learned and experienced standard-based, performance-oriented and battle-focused training in support of team level operations. During this course, combat leadership skills were trained and reinforced.
PLDC	
NCO Academy	
Ft. Polk, LA	

PROFESSIONAL EXPERIENCE

LAW ENFORCEMENT

U.S. Marshals	Kansas City, MO	2008-Present

Supervisory Criminal Investigator-Deputy U.S. Marshal GS-1811

Supervisor of district operations.

Provide national tracker certification and re-certification.

Provide hoplology analysis: self-defense and tracking forensic/crime scene specialty, and expert testimony/consultation.

U.S. Marshals	Kansas City, MO	2007-2008

Criminal Investigator-Deputy U.S. Marshals GS-1811

Worked protective operations of U.S. Supreme Court Justices, U.S. District Court Judges and other National and Foreign Dignitaries.

Worked court operations.

Provide national tracker certification and re-certification.

Provide hoplology analysis: self-defense and tracking forensic/crime scene specialty, and expert testimony/consultation.

U.S. Marshals	Cheyenne, WY	2002-2007

Supervisory Criminal Investigator-Deputy U.S. Marshal GS-1811

& Chief of Scouts

Developed and led the Wyoming U.S. Marshals Mounted Tracking Unit (MTU)

Provide national tracker certification and re-certification.

Provide hoplology analysis: self-defense and tracking forensic/crime scene specialty, and expert testimony/consultation.

Supervise and teach tactical firearms employment and applications for MTU

Supervisor of the Wyoming Fugitive Task Force.

Supervisor of the Operational Mission (Court, Warrants, Threat) of the District of Wyoming.

Supervisor of district protective operations.

U.S. Marshals   Anchorage, AK   1991-2002
Criminal Investigator-Deputy U.S. Marshals GS-1811

Worked protective operations of U.S. Supreme Court Justices,
    U.S. District Court Judges, and other National and Foreign Dignitaries.
Worked court operations and fugitive warrant investigations.
    Developed and Led the Alaska U.S. Marshals Tactical Tracking Unit (TTU).
Provide national tracker certification and re-certification.
Provide hoplology analysis: self-defense and tracking forensic/crime scene specialty,
    and expert testimony/consultation.
Team operator on the Anchorage Regional FBI SWAT Team.

MILITARY

4th ID (ROC), WYARNG   Casper, WY   2005-2006
Assistant Operations Sergeant
- Employment of Infantry small unit tactics.
- Training and maintenance of subordinate soldiers.
- Integration with staff officers to accomplish operational objectives at the staff level.
- Terrain management.
- Development of tactical courses of action.
- Employ the military decision-making process for strategic and tactical applications.

TSC, WYARNG   Camp Guernsey, WY   2003-2005
Range Control Specialist

Liaison with units utilizing the ranges.
Training units in small arms employment (M9, M16, M24, M249, M60).
Train units in combat man-tracking
Maintenance of ranges.

Det 2, LRSD, 207th, AKARNG   Camp Carroll, AK   1988-1997
Team Sergeant, Senior Scout, Sniper, Tracker
- Reconnaissance and Surveillance.
- Target Interdiction.
- Screen operations.
- Airborne operations.
- Arctic operations.

Active Duty   1984-1988
Infantryman, Team Leader, Squad Leader, Sniper
Alpha CO., 1/325th Inf., 82nd Airborne Division, Ft. Bragg, N.C.
Alpha CO., 1/ 61st., Inf., 5th Infantry Division, Ft. Polk, LA
Charlie CO., 1/17th Inf., 6th Infantry Division, Ft. Richardson, AK

- Airborne Infantry-Combat Arms and Sniper Operations.
- Use small arms weapons and combat equipment.
- Responsible for team or squad.

TRACKING EXPERIENCE

International Society of Professional Trackers   AK, WY, MO   1997-Present
Founder, Tracking Historian, & Staff Writer
Staff writer for Track & Sign "Founder's Track Trap" series.
Tracking Historian for the ISPT Archives and Library of Tracking
    Research.

International Society of Professional Trackers   Wasilla, Alaska   1997-1999
Founder & Executive Director
Man and animal tracking services to all trackers worldwide.
Established worldwide organization of trackers.
Organized ISPT objectives and instituted first international symposium.

Lost Trail Ranch & Institute-Scout Tracking, Survival, Defense School   1988-Present
Director & Chief Tracker
Worked man and animal tracking services for federal, state, and
    local government.
Teaching of man, animal, tactical, combat, search & rescue,
    law enforcement, crime scene tracking skills and forensic development.
Research in the areas of footfall trace and impression evidence;
    urban, wilderness, animal crime scene management.

CURRENT CERTIFICATIONS, QUALIFICATIONS AND LICENSURE

Tracking Specialty Certifications
(U.S. Marshals Service-District of Alaska & Wyoming):

Mounted Backcountry	2005
Pack Handler	2005
Wilderness & Animal Crime Scene	2001
Urban Crime Scene	2001
Instructor	2001
Search and Rescue	1997
Tactical	1995
Reconnaissance & Surveillance	1993
Certified Master Scout Tracker, USMS-D/WY	2004
Supervisory Criminal Investigator-Deputy U.S. Marshal USMS-D/WY	2002
Certified Senior Scout Tracker, USMS-D/AK	2002
Certified Scout Tracker, USMS-D/AK	1999
Certified Police Instructor, APSC	1999

(Tracking and Defensive Tactics)

Certified Scout, USMS-D/AK	1997
SWAT Operator, FBI-Anchorage	1996
Criminal Investigator-Deputy U.S. Marshal, USMS-D/AK	1991
Personal Protection Specialist, USMS-D/AK	1991
U.S. Army Sniper, U.S. Army	1987
U.S. Army Arctic Survival Specialist, U.S. Army	1987
Advanced Parachutist	1984
Basic Parachutist	1984

CURRENT CLEARANCES

Top Secret	U.S. Marshals	2003-Present
Secret	U.S. Marshals	1991-2003
Secret	U.S. Army Reserves & Guard	1988-2006
Secret	U.S. Army	1984-1988

HONORS & AWARDS

United States Marshals Service Special Service Award	2008
United States Marshals Service Special Service Award	2007
United States Marshals Service Outstanding Performance Award	2000

Letter of Outstanding Performance, Director of U.S. Marshals Service
    Awarded for recognition of outstanding performance of official
    duties by providing protection of Elian Gonzalez and his father
    during Operation Quincy.
	2000

Letter of Outstanding Performance, Director of U.S. Marshals Service
    Awarded for recognition of outstanding performance of official
    duties by teaching prevention of workplace violence to the
    Wrangell Police Dept.
	2000

United States Marshals Service Superior Performance Award
    Awarded for superior performance of official duties by teaching
    defensive tactics and man-tracking to the military and law
    enforcement communities.
	1997

RECOGNITIONS & APPEARANCES

BOOKS

"The Hell Riders", Pinnacle Books	New York, NY	2004
"Hard Road to Heaven", Pinnacle Books	New York, NY	2002
"Hide & Seek", Publication Consultants	Anchorage, Alaska	2000
"Pray for Justice", Publication Consultants	Anchorage, Alaska	1998

RADIO

"Ty's Tuesday Tracking Tips" Alaska Outdoor Magazine Radio Show, KBYR 700	Anchorage, AK	1999-2000
"Tracking in Law Enforcement", KBYR 700	Anchorage, AK	1997

TELEVISION

"Predator Instinct" Red Brick Entertainment	New York, NY	2006
"The Chase: Track Down", Court TV	New York, NY	2005
"Women in Law Enforcement," Oxygen TV Network	Anch., AK; NY, NY	2000

STATE

The Skamania County Pioneer	Stevenson, WA	1999
Anchorage Daily News	Anchorage, AK	1998

NATIONAL

The Frederick Russell Burnham Historical Society	Tucson, AZ	2004
U.S. Fish and Wildlife Newsletter	World Wide Web	2000
National Native American Law Enforcement Officers Association Newsletter	Washington, D.C	1998
U.S. Marshals Service Newsletter	Arlington, VA	1998

INTERNATIONAL

Man Tracking University	World Wide Web	2007
Royal Canadian Mounted Police Magazine, "Gazette" – Modern Day Scouts	Quebec, Canada	1999
International Society of Professional Trackers Track and Sign Newsletter	Santa Rosa, CA	1998-Present

APPLIED RESEARCH: SPOOROLOGY/MANTRACKING-TECHNICAL & FORENSICS ADVANCEMENTS

Behavior, Gait, Spoor (BGS) Paradigm
Observation, Classification, Interpretation of Spoor Evidence (OCISE) Algorithm
Spoor Platting
Spoor-Chain Signature
Classification of Spoor
Foot Measurement Analysis
Quadrant Reference Analysis
Ground Contact Points
Gait Platting
Gait-Footfall Sequencing
Gait Biometric Identification
Behavior Platting

Aerial Travel Points
Cone of Entry-Cone of Exit
Pre-Trailing Checklist
Initial Trail Assessment
Trail Confirmation Standard
Post-Trailing Checklist
Target Quarry Reference
Mechanical Stride Zero
Tracker's Triangle
Trails Leading Edge
Human Behavior Identification
Deliberate Trailing Method
Dynamic Trailing Method
Foot Impression Reference Matrix
Foot Roll Analysis (Impact Point Gradient, Impact Point, Flex Point, Pivot Point, Terminal Point, Terminal Point Gradient)
Apex Stride Step Estimate
Variable Trail Patterns
Ground Hardness Type
Ground Surface Type
Track Erosion Computation
Register Deviation Anomaly
Minimum Track Report
Slope Topography Analysis
Tracker Sketch Analysis
Uniform Scale Format
Eight Rules of Tracking Forensics
Tracking Crime Scene Format
Spoor Deterioration Risk Factors
Spoor Models of Interpretation & Projection
Seven Tracking Forensic Skills Steps
Tracking Communication Control Alpha-Numerics
Blood Spoor Forensics
Forensic Spoor Description
Crime Scene Tracking Kit
Trailing Profile Checklist-Man Tracking Blueprint
Forensic Trailing Outline
Scout Tracker Certification
Forensic Spoor Card
Quarry Spoor Formula

CASE EXPERTISE AND SPECIAL TESTIMONY

U.S. District Judge, The Honorable John Sedwick, Utilities tampering case, U.S. District Court, Anchorage, Alaska
Clinton Baker Murder Case, Alaska Fugitive Task Force, Anchorage, AK
Amy and Lydia Byron Missing Person Case, Alaska State Troopers, Palmer, AK
Susan Beth Crawford Missing Person Case, Alaska State Troopers, Glennallen, AK
Wounded & Missing Brown Bear Recovery Case, ADFG, Ft. Richardson, AK
Nathan Swain Domestic Violence/foot pursuit Case, CPD, Cheyenne, WY
Matthew Davis Multiple Burglary Case, CPD, Cheyenne, WY
Ellory McCaulley Missing Person (Drowning) Case, Carbon County, Rawlins, WY
Evidence Search for a Missing Weapon Case-Uinta County, Evanston, WY
Obsidian Theft Case-Yellowstone National Park, U.S. Park Service
U.S. District Judge, The Honorable Ortrie Smith, Burglary/ Stolen government property case, Kansas City, Missouri

CONSULTING AND TRAINING

Juneau Police Department	Juneau, AK
Alaska State Troopers	Alaska Statewide
Fish & Game Troopers	Alaska Statewide
Anchorage Police Department	Anchorage, AK
Wasilla Police Department	Wasilla, AK
Palmer Police Department	Palmer, AK
U.S. Marshals Service	Nationwide
U.S. District & Magistrate Court	Nationwide
Alaska Department of Corrections	Alaska Statewide
U.S. Immigration and Naturalization	Anchorage, AK
U.S. Drug Enforcement Administration	Anchorage, AK
U.S. Alcohol, Tobacco, Firearms, and Explosives	Anchorage, AK
U.S. Federal Bureau of Investigation	Anchorage, AK; Lander, WY
U.S. Secret Service	Anchorage, AK; Cheyenne, WY
National Park Service	Glacier Bay National Park, AK
Chugach State Park	Anchorage, AK
U.S. Bureau of Land Management	Fairbanks, AK
Chickaloon Village Traditional Council	Chickaloon, AK
Grand County Sheriff's Office SAR	Moab, UT
Moab Police Department	Moab, UT
Wind River Indian Reservation Police	WRIR, WY
Wind River Indian Reservation Fish and Game Office	WRIR, WY
U.S. Bureau of Indian Affairs	WRIR, WY
Wyoming Highway Patrol	Wyoming Statewide
Logan County Sheriff's Office	Sterling, CO

179

Washington County Sheriff's Office	Akron, CO
Fremont County Sheriff's Office	Riverton, WY
Riverton Police Department	Riverton, WY
Hot Springs County Sheriff's Office	Thermopolis, WY
Lincoln Police Department	Lincoln, NE
Uinta County Sheriff's Office	Evanston, WY
Lancaster County Sheriff's Office	Lincoln, NE
U.S. Army SF, Rangers, Airborne, Inf.	Nationwide
Alaska Army National Guard	Alaska Statewide
Alaska Air Guard, Kulis Air Base	Anchorage, AK
Yukon Quest Sled Dog Race	Fairbanks, AK
U.S. Fish and Wildlife Service	Anchorage, AK; Fairbanks, AK; Juneau, AK; Lander, WY
U.S. Forest Service	Tongass National Forest, AK; Chugach National Forest, AK; Shoshone National Forest, WY; Bridger-Teton National Forest, WY

Consultation given also to public and private groups, programs, and task forces throughout the U.S. and Canada. Keynote speaker at international conferences and gatherings.

PROFESSIONAL ASSOCIATIONS

International Hoplology Association	2010
Federal Law Enforcement Officers Association	
Wyoming Chapter	2002-2007
Alaska Chapter	1991-2002
International Society of Professional Trackers (Life Member)	
Tracking Historian, ALTR	2007
Editor, The Founder's Track Trap, Track & Sign	2003
Charter Member	1998
Co-founder	1997
The Frederick Russell Burnham Historical Society of Tucson	
Charter member	2004

TEACHING EXPERIENCE

Chief Tracker, U.S. Marshals Mounted Tracking Unit Scout Tracking Operations United States Marshals Service-District of Wyoming	2003-2006
Chief Tracker, U. S. Marshals Tactical Tracking Unit Scout Tracking Operations United States Marshals Service-District of Alaska	1998-2002

Adjunct Instructor, Universal Tracking Services   1997-1998
Everson, WA

Chief Instructor, Police Ju-Jitsu (Defensive Tactics), Officer   1994-2002
Survival, and VIP Protection
United States Marshals Service-District of Alaska, Anchorage, AK

SPECIALTY TRAINING COURSES AND SCHOOLS

Basic Backcountry Mounted Enforcement School,   2005
40 hours, Powell, Wyoming

Improvised Explosive Devices, 16 Hours   2004
10th Special Forces, Camp Guernsey, WY

High Arctic Survival, 8 hours   2001
Anchorage, AK

Shoe and Tire Evidence Specialist Course, 8 hours   2000
Anchorage, AK

Basic Meteorology, Cloud Identification and Forecasting   2000
8 hours, Elmendorf AFB, AK

LE, MIL, SAR Tracking Seminar, 24 hours   1999
Petaluma, CA

Arctic and Sub-Arctic Survival Training, (Summer)   1998
72 Hours, Talkeetna, AK

Advanced Deputy U.S. Marshals Course, 72 hours   1998
FLETC, Glynco, GA

Methods of Instruction Course, 40 hours   1998
Alaska State Troopers, Ft. Richardson, AK

Cold Weather Survival, 8 hours   1995
FBI, Ft. Richardson, AK

SWAT Training Course, 56 hours   1995
FBI, Salt Lake City, UT

Advanced Deputy U.S. Marshals Course, 80 hours   1995
FLETC, Glynco, GA

Long-Range Surveillance Course, 120 hours   1993
Schofield Barracks, Oahu, Hawaii

Basic Deputy U.S. Marshals Course, 200 hours   1991
FLETC, Glynco, GA

Criminal Investigator Course CI-114, 340 Hours   1991
FLETC, Glynco, GA

PUBLICATIONS

Cunningham, T.J. (2009, Summer). The founder's track trap: Scout tracker team mission statement and policy go-by (Part One). *Track & Sign*. Santa Rosa, CA: ISPT.

Cunningham, T.J. (2008, Spring). The founder's track trap: Scout tracker training programs. *Track & Sign, 28,* 2-5. Santa Rosa, CA: ISPT.

Cunningham, T.J. (2007, Winter). The founder's track trap: The face of scout tracking. *Track & Sign, 27,* 2-3. Santa Rosa, CA: ISPT.

Cunningham, T.J. (2007, Summer). The founder's track trap: Weather, aging spoor, and my TEC formula. *Track & Sign, 25,* 2-13. Santa Rosa, CA: ISPT.

Cunningham, T.J., Rowe, C., and Otte, M. (2006). *Use of Mules and Horses in the Backcountry: Packing and Tracking.* Cheyenne, WY: U.S. Marshals Service-District of Wyoming: Mounted Tracking Unit.

Cunningham, T.J. (2006). *Scout tracking: Silent team signals during visibility conditions.* Cheyenne, WY: U.S. Marshals Service-District of Wyoming: Mounted Tracking Unit.

Cunningham, T.J., and Steele, B. (2006). *Working at night for law enforcement and military.* Cheyenne, WY: U.S. Marshals Service-District of Wyoming: Mounted Tracking Unit.

Cunningham, T.J., and Steele, B. (2006). *Geographic considerations for law enforcement and military.* Cheyenne, WY: U.S. Marshals Service-District of Wyoming: Mounted Tracking Unit.

Cunningham, T.J. (2006, Winter). The founder's track trap: Staining. *Track & Sign, 22,* 2-3. Santa Rosa, CA: ISPT.

Cunningham, T.J. (2005). *Crime scene tracking for law enforcement and military.* Cheyenne, WY: U.S. Marshals Service-District of Wyoming: Mounted Tracking Unit.

Cunningham, T.J. (2005). *Search and rescue tracking for law enforcement and military.* Cheyenne, WY: U.S. Marshals Service-District of Wyoming: Mounted Tracking Unit.

Cunningham, T.J. (2003). Methods of observation. *Wilderness Way, 10*(4), 25-28.

Cunningham, T.J. (2005, Summer). The founder's track trap: Scout tracking published. *Track & Sign, 21,* 2. Santa Rosa, CA: ISPT.

Cunningham, T.J., and Orde, C. (2005). *Use of Tracking Dogs for Law Enforcement, Military, and Search & Rescue.* Cheyenne, WY: U.S. Marshals Service-District of Wyoming: Mounted Tracking Unit.

Cunningham, T.J. (2005, Winter). The founder's track trap: Scout tracker virtues (Part VII). *Track & Sign, 20,* 2-3. Santa Rosa, CA: ISPT.

Cunningham, T.J. (2003). What to look for: An introduction to tracking. *Wilderness Way, 10*(1), 23-27.

Cunningham, T.J. (2004). *Scout tracking operations (Vol. 2): Basic trailing and surveillance concepts for operational trackers.* WY: Lost Trail Institute.

Cunningham, T.J. (2004). *Scout tracking operations (Vol. 1): Basic trailing and surveillance concepts for operational trackers.* WY: Lost Trail Institute.

Cunningham, T.J. (2004, Spring). The founder's track trap: Scout tracker virtues (Part VI). *Track & Sign, 19,* 2. Santa Rosa, CA: ISPT.

Cunningham, T.J. (2004, Spring). The founder's track trap: Scout tracker virtues (Part V). *Track & Sign, 18,* 2. Santa Rosa, CA: ISPT.

Cunningham, T.J. (2004). *Scout tracking: How to see and what to look for.* Cheyenne, WY: U.S. Marshals Service-District of Wyoming: Mounted Tracking Unit.

Cunningham, T.J. (2004, Winter). The founder's track trap: Scout tracker virtues (Part IV). *Track & Sign, 17,* 2. Santa Rosa, CA: ISPT.

Cunningham, T.J. (2003). How to see: An introduction to tracking. *Wilderness Way, 9*(3), 6-10.

Cunningham, T.J. (2003, Spring). Master tracker series: Understanding the seven steps using the key word T.R.A.C.K.E.R.-part two. *Primitive Archer, 11*(1), 51-67.

Cunningham, T.J., Otte, M., and Sands, H. (2003). *Scout survival for law enforcement and military.* Cheyenne, WY: U.S. Marshals Service-District of Wyoming: Mounted Tracking Unit.

Cunningham, T.J. (2003, Autumn). The founder's track trap: Scout tracker virtues (Part III). *Track & Sign, 16,* 2. Santa Rosa, CA: ISPT.

Cunningham, T.J., and Otte, M. (2003). *Scout tracker trailing outline.* Cheyenne, WY: U.S. Marshals Service-District of Wyoming: Mounted Tracking Unit.

Cunningham, T.J. (2003, Spring/Summer). The founder's track trap: Scout tracker virtues (Part II). *Track & Sign, 15,* 3. Santa Rosa, CA: ISPT.

Cunningham, T.J., and Guinn, K. (2003). *Surveillance and reconnaissance for law enforcement and military.* Cheyenne, WY: U.S. Marshals Service-District of Wyoming: Mounted Tracking Unit.

Cunningham, T.J., and Guinn, K. (2003). *Surveillance hides for law enforcement and military.* Cheyenne, WY: U.S. Marshals Service-District of Wyoming: Mounted Tracking Unit.

Cunningham, T.J. (2003, Winter). The founder's track trap: Scout tracker virtues (Part I). *Track & Sign, 14,* 2-11. Santa Rosa, CA: ISPT.

Cunningham, T.J. (2002, Winter). Master tracker series: Understanding the seven steps using the key word T.R.A.C.K.E.R.. *Primitive Archer, 10*(4), 60-67.

Cunningham, T.J. (2002, Fall). Master tracker series: Eight rules for tracking an animal. *Primitive Archer, 10*(3), 60-66.

Cunningham, T.J. (2002, Summer). Master tracker series: Using your bow to trail an animal. *Primitive Archer, 10*(2), 16-25.

Cunningham, T.J. (2002, March). The trailing mission and your sphere of influence. *Track & Sign, 12,* 10-13. Santa Rosa, CA: ISPT.

Cunningham, T.J. (2001). *Principles and virtues of the scout tracker.* Anchorage, AK: U.S. Marshals Service-District of Alaska: Tactical Tracking Unit.

Cunningham, T.J. (2001). *Six basic indicators of scout tracking.* Anchorage, AK: U.S. Marshals Service-District of Alaska: Tactical Tracking Unit.

Otte, M., and Cunningham, T.J. (2000). *Learning to track Alaska's wildlife: Activity coloring book.* Eagle River: Broken Heart Books.

Cunningham, T.J. (2000). *The eight rules of T.R.A.C.K.I.N.G.* Anchorage, AK: U.S. Marshals Service-District of Alaska: Tactical Tracking Unit.

Cunningham, T.J., and Guinn, K. (1999). *Scout tracker team mission statement and policy go-by.* Anchorage, AK: U.S. Marshals Service-District of Alaska: Tactical Tracking Unit.

Cunningham, T.J., Otte, M., Phillips, W., Johnson, R., Murphy, J. (1999). *Scout Tracker Certification Procedures for Law Enforcement, Military, and Search & Rescue.* Anchorage, AK: U.S. Marshals Service-District of Alaska: Tactical Tracking Unit.

Cunningham, T.J. and Otte, M. (Eds.). (1999). *Scout tracker training program.* Anchorage,AK: U.S. Marshals Service-District of Alaska: Tactical Tracking Unit.

Cunningham, T.J. (1999). *Scout tracking: Comprehensive glossary of terms.* Anchorage, AK: U.S. Marshals Service-District of Alaska: Tactical Tracking Unit.

Cunningham, T.J., and Guinn, K. (1998). *Scout tracker training program.* Anchorage, AK:U.S. Marshals Service-District of Alaska: Tactical Tracking Unit.

Cunningham, T.J. (1998). *The scout tracker: Responding to the trail.* Anchorage, AK: U.S. Marshals Service-District of Alaska: Tactical Tracking Unit.

Cunningham, T.J. (1998). *The scout tracker: Enlarging the trail.* Anchorage, AK: U.S. Marshals Service-District of Alaska: Tactical Tracking Unit.

Cunningham, T.J. (1998). *The scout tracker: Knowing the trail.* Anchorage, AK: U.S. Marshals Service-District of Alaska: Tactical Tracking Unit.

Cunningham, T.J. (1998). *The scout tracker: Confirming the trail.* Anchorage, AK: U.S. Marshals Service-District of Alaska: Tactical Tracking Unit.

Cunningham, T.J. (1998). *The scout tracker: Assessing the trail.* Anchorage, AK: U.S. Marshals Service-District of Alaska: Tactical Tracking Unit.

Cunningham, T.J. (1998). *The scout tracker: Receiving the trail.* Anchorage, AK: U.S. Marshals Service-District of Alaska: Tactical Tracking Unit.

Cunningham, T.J. (1998). *The scout tracker: Thinking of the trail.* Anchorage, AK: U.S. Marshals Service-District of Alaska: Tactical Tracking Unit.

KEYNOTE SPEECHES, LECTURES, PRESENTATIONS, AND COURSES

State/Regional:

Cunningham, T.J. (2007). *Scout tracking operations-basic course-60 hours.* Invited in Service Training for the U.S. Marshals Tactical Tracking Unit, District of Alaska, Mulchatna River, AK., July 2007.

Cunningham, T.J. (2005). *Law enforcement tracking.* Invited Seminar for the Lancaster County Sheriff's Office, Lincoln, NE, 2005.

Cunningham, T.J., and Bort, B. (2005). *Search and rescue tracking.* Invited Course for the Lincoln County SAR Group, Star Valley, WY, 2005.

Cunningham, T.J., Mears, J., Von Rein, C. (2005). *Law enforcement tracking.* Invited Course for the Logan County Sheriff's Office, Sterling, CO, 2005.

Cunningham, T.J., Von Rein, C., Rose, T. (2005). *Search and rescue tracking.* Invited Course for the Grand County SAR Group, Moab, UT, 2005.

Henderman, D., Cunningham, T.J., Speiden, R., Cire, A., Erchak, I. (2004). *Basic tracking principles.* Invited presentation at the Court TV production of The Chase: Trackdown, On-location, Pocono, PA, October 2004.

Cunningham, T.J., Tippy, B, Rager, S., Kirkman, B. (2004). *Scout tracking basic course.* Invited Course for Federal Bureau of Investigation, Ft. Wasakie, WY, June 2004.

Tippy, B, Rager, S., Kirkman, B., Cunningham, T.J. (2004). *Law Enforcement Tracking.* Invited Course for Wyoming Highway Patrol, Camp Guernsey, WY, April 2004.

Tippy, B, Rager, S., Kirkman, B., Cunningham, T.J. (2004). *Law Enforcement Tracking.* Invited Course for Wyoming Highway Patrol, Camp Guernsey, WY, April 2004.

Tippy, B, Rager, S., Kirkman, B., Cunningham, T.J. (2004). *Law Enforcement Tracking.* Invited Course for Wyoming Highway Patrol, Camp Guernsey, WY, March 2004.

Tippy, B, Rager, S., Kirkman, B., Cunningham, T.J. (2004). *Law Enforcement Tracking.* Invited Course for Wyoming Highway Patrol, Camp Guernsey, WY, March 2004.

Cunningham, T.J. (2002). *Scout tracking extended operations-60 hours.* In-service training course for the U.S. Marshals Tactical Tracking Unit, District of Alaska, Prince William Sound, AK. May 2002.

Cunningham, T.J., Otte, M., and Sands, H. (2002). *Arctic survival, snow tracking and Movement-8 hours.* In-service training course given to the U.S. Marshals Tactical Tracking Unit, District of Alaska, Fairbanks, AK, February 2002.

Cunningham, T.J. (2001). *Scout tracking instructors course-200 hours.* In-service training course given incrementally from March 7, 1997 through May 25, 2001, U.S. Marshals Tactical Tracking Unit, Anchorage, Alaska, May 2001.

Cunningham, T.J. (2001). *Scout Tracking Advanced Course-56 hours.* In-service training course given to the U.S. Marshals Tactical Tracking Unit, Anchorage, Alaska, May 2001.

Cunningham, T.J., and Otte, M. (2001). *Arctic Survival, Terrain Analysis- 8 hours.* In-service Training course given to the U.S. Marshals Tactical Tracking Unit, Ft. Richardson, Alaska, April 2001.

Cunningham, T.J. (2001). *Improvised Explosive Devices, Anti-Tracking, Reduction and Counter Tracking-8 hours.* In-service training course given to the U.S. Marshals Tactical Tracking Unit, Ft. Richardson, Alaska, March 2001.

Phillips, W., Cunningham, T.J., Otte, M., Guinn, K. (1998). *Basic tracking principles.* Invited presentation at Oxygen Television Network, On-location, Anchorage, AK, September 2000.

Cunningham, T.J., and Guinn, K. (2000). *Surveillance/observation, hides, use of optics-8 hours.* In-service training course given to the U.S. Marshals Tactical Tracking Unit, Ft. Richardson, Alaska, June 2000.

Cunningham, T.J. (2000). *Scout tracking basic course-56 hours.* In-service training course given to the U.S. Marshals Tactical Tracking Unit, Susitna, AK, May 2000.

Hardin, J., Cunningham, T.J., and Others. (1998). *SAR Tracking Specialist Course-24 Hours.* Invited course for Search and Rescue Groups around Henry Coe State Park, California, August 1998.

Hardin, J., Cunningham, T.J., and Others. (1997). *Basic Tracking Course-24 Hours.* Invited course for Alaskan Law Enforcement and SAR Groups, Camp Denali, AK, July 1997.

Hardin, J., Cunningham, T.J., and Others. (1997). *Basic Tracking Course-24 Hours.* Invited course for Idaho Law Enforcement and SAR Groups, McCall, Idaho, June 1997.

Hardin, J., Cunningham, T.J., and Others. (1997). *Basic Tracking Course-24 Hours.* Invited course for Washington Law Enforcement and SAR Groups, Washington County, WA, April 1997.

Hardin, J., Cunningham, T.J., and Others. (1997). *Basic Tracking Course-24 Hours.* Invited course for Idaho Law Enforcement and SAR Groups, Blaine County, Idaho, March 1997.

National:

Hardin, J., Larue, R., Cunningham, T.J., and Others. (1998). *Basic combat man-tracking.* Invited presentation at U.S. Army 7th Special Forces, Ft. Bragg, N.C., June 1998.

International:

Cunningham, T.J. (2004). *The tracker's triangle.* Invited lecture at the International Society of Professional Trackers Symposium, Appomattox, VA, October 2004.

Cunningham, T.J. (1998). *Professional neutrality in analytical application of the science of tracking.* Invited lecture at the International Society of Professional Trackers Symposium, Petaluma, CA, October 1998.

BOOKS AND PAPERS REVIEWED-TRACKING, SURVIVAL, DEFENSE TECHNICAL ADVISOR

Song, T. (2010). Song of the Track: Adventures and Instructions in the Old Native Way of Tracking.	2010
Speiden, R. (2008). Foundations for awareness, Signcutting and Tracking. VA: Natural Awareness Publishing.	2008
Henry, M. (2004). The Hellriders. NY: Pinnacle Books.	2006
Henry, M. (2002). Hard Road to Heaven. NY: Pinnacle Books.	2005
Otte, M. (1999). Hide and Seek. AK: Publication Consultants.	1999
Otte, M. (1998). Pray for Justice. AK: Publication Consultants.	1997

NATIONAL SCOUT-TRACKERS CERTIFIED

Over 100

BIBLIOGRAPHY

Books

Bodziak, William. *Footwear Impression Evidence: Detection, Recovery, and Examination.* Boca Raton, FL: CRC Press, 2000.

Brown, Tom Jr., and Brandt Morgan. *Tom Brown's Field Guide to Nature Observation and Tracking.* New York: Berkley Books, 1983.

Brown, Tom Jr. 1999. *The Science and Art of Tracking: Nature's Path to Spiritual Discovery.* New York: Berkley Books.

Carss, Bob. *The SAS Guide to Tracking.* New York: The Lyons Press, 2000.

Dispenza, Joe. *Evolve Your Brain: The Science of Changing your Mind.* Deerfield Beach, FL: Health Communications, 2007.

Gilbert, Adrian. *Sniper.* New York: St. Martin's Press, 1994.

Hardin, Joel, and Matt Condon. *Tracker: Case Files and Adventures of a Professional Mantracker.* Everson, WA: Joel Hardin Professional Tracking Services, 2007.

Hurth, John D., and Jason W. Brokaw. *Visual Tracking and the Military Tracking Team Capability: A Disappearing Skill and Misunderstood Capability.* Carencro: TYR Group, LLC, 2010.

James, William. *The Principles of Psychology.* Vol. 2. New York: Dover Publications, 1950.

Kearney, Jack. *Tracking: A Blueprint for Learning How.* El Cajon: Pathways Press, 1996.

Kuhn, Thomas. *The Structure of Scientific Revolutions,* 2d ed. Chicago, IL: University of Chicago Press, 1962.

Liebenberg, Louis. *The Art of Tracking: The Origin of Science.* Cape Town: David Phillips, 1990.

Lee, Lily. *Gait Dynamics for Recognition and Classification.* Cambridge: Massachusetts Institute of Technology, 2001.

Robbins, Roland. *Mantracking: Introduction to the Step-by-Step Method.* Montrose: Search and Rescue Magazine, 1977.

Saferstein, Richard. *Criminalistics: An Introduction to Forensic Science.* New Jersey: Pearson Education, 2007.

Scott-Donelan, D. *Tactical Tracking Operations: The Essential Guide for Military and Police Trackers*. Boulder, CO: Paladin Press, 1998.

Speiden, Robert. *Foundations for Awareness, Signcutting and Tracking*. Christianburg, VA: Natural Awareness Tracking School, 2009.

Taylor, Albert, and Donald Cooper. *Fundamentals of Mantracking: The Step-by Step Method*. Olympia: ERI International, 1990.

Thomas-Cottingham, Alison. *Psychology Made Easy*. New York: Broadway Books, 2004.

Government Documents

Headquarters, Department of the Army. ATTP 3-39.20, *Police Intelligence Operations*. Washington, DC: Government Printing Office, 2010.

U.S. Marshals Service. "Scout Tracking Basic Course." Fort Richardson and Susitna, AK, 2000.

———. *Scout Tracking Operations Course: Law Enforcement-Certified Tracker I, II, III*. Anchorage: District of Alaska Tactical Tracking Unit, 2000.

Periodicals

Armstrong, Hunter B. "Pre-Arranged Movement Patterns." *HOPLOS* 4, no.1 and 2. (Winter 1988) Palm Desert: International Hoplology Society.

De Sousa, Ronald. "Emotion." *The Stanford Encyclopedia of Philosophy* (Spring 2010), ed. Edward N. Zalta. http://plato.stanford.edu/archives/ spr2010/entries/emotion/ (accessed 6 April 2011.

Hanratty. Tom. "Sherlock Holmes, Master Tracker, Part 1." *Track & Sign* 29 (2008).

Hilderbrand, Dwayne S. "Footwear, The Missed Evidence." http://www.crime-scene-investigator.net/footwear.html (accessed 22 August 2010).

Hull, Michael. "Tracking as evidence." *Track & Sign* 28 (2008).

Sacks, Dave. 1998. "Tracking." *Gazette* 61 (February/March 1999).

<u>Other Sources</u>

Ayyappa, Ed. "Normal Human Locomotion, Part 1: Basic Concepts and Terminology." *American Academy of Orthotists & Prosthetists* 99, no.1 (1997). http://www.oandp.org/jpo/library/1997_01_010.asp. (accessed 23 October 2009).

Beckerman, Linda P. "Non-Linear Dynamics of War." http://www.calresco.org/beckerman/nonlindy.htm (accessed 5 April 2011).

Berger, Kevin E. *Tracking Report*. Virginia Department of Game and Inland Fisheries. 7 July 1999.

Black, Catherine M. "Legal Implications of the Use of Biometrics as a Tool to Fight the Global War on Terrorism." Master's Thesis, U.S. Army Command and General Staff College, 2008.

Cornell University Law School. *Federal Rules of Evidence.* http://www.law.cornell.edu/rules/fre/rules.htm#Rule702 (accessed 5 April 2011).

———. *Kumho Tire Co., Ltd. v. Carmichael.* http://www.law.cornell.edu/supct/html/97-1709.ZO.html (accessed 5 April 2011).

Cunningham, T.J. *Scout Tracking Operations: Basic Trailing and Surveillance Concepts for Operational Trackers*. Cheyenne, WY: Lost Trail Institute, 2004.

Fautua, David, Sae Schatz, David Kobus, V. Alan Spiker, William Ross, Joan H. Johnston, Denise Nicholson, and Emilie A. Reitz. *Border Hunter Research Technical Report*. Norfolk, VA: U.S. Joint Forces Command, 2010.

Findlaw for Legal Professionals, *Daubert v. Merrill Dow Pharmaceuticals, Inc.*, http://caselaw.lp.findlaw.com/cgi-bin/getcase.pl?court=US&vol=509&invol=579 (accessed 5 April 2011).

Gyllensporre, Dennis T. "Adding Nonlinear Tools to the Strategist's Toolbox." Master's Thesis, U.S. Army Command and General Staff College, 2001.

Hansen, Mark. "He Tries Men's Soles." *American Bar Association.* http://www.aba journal.com/magazine/article/he_tries_mens_soles/ (accessed 17 September 2010).

Hayes, Richard. *Paleolithic Adaptive Traits and the Fighting Man.* Sedona, AZ: International Hoplology Society, 1998.

Holistic Online.com. "Stress: The Silent Killer." http://holisticonline.com/stress/stress_gas.htm (accessed 6 April 2011).

Huitt, W. "A Systems Approach to the Study of Human Behavior." *Educational Psychology Interactive.* Valdosta, GA: Valdosta State University. http://www.edpsycinteractive.org/materials/sysmdlo.html (accessed 16 September 2010).

Kale, A., N. Cuntoor, B. Yegnanarayana, A.N. Rajagopalan, and R. Chellappa. "Gait Analysis for Human Identification." http://www.cfar.umd.edu/~cuntoor/kale03gait.pdf (accessed 19 November 2010).

Nowlan, Michael. "Human Recognition via Gait Identification Using Accelerometer Gyro Forces." http://cs-www.cs.yale.edu/homes/mfu3/proj/mfu_gait_id.pdf (accessed 13 July 2010).

Reeve, Kevin. *Tracking.* Vincentown: On Point Tactical Tracking Services.

Southwest Guidebooks. "The Four Corners Fugitive Search- the Largest Manhunt in Western History." http://www.southwestguidebooks.com/fugitives.htm (accessed 24 July 2010).

State of Alaska. *Homicide, 98-8267*, by Craig Allen. Alaska State Troopers. Glennallen. 31 January 1998.

Thorton, Ian, Jeannine Pinto, Maggie Shiffrar. "The Visual Perception of Human Locomotion." *Cognitive Neuropsychology* 15, 6/7/8 (1998). http://psychology.rutgers.edu/~mag/reprint_pdfs/TPS98.pdf (accessed 19 September 2010).

Vallandigham, Paul. "Tracking." Paper presented at the Annual Symposium of the International Society of Professional Trackers, Petaluma: CA. 22-24 October 1999.

———. "Tracking: 19th Century Art as 21st Century Science." Paper presented at the Annual Symposium of the International Society of Professional Trackers, Petaluma: CA. 22-24 October 1999.

———. "Eye Dominance and Your Body." Paper presented at the Annual Symposium of the International Society of Professional Trackers, Petaluma: CA. 22-24 October 1999.

INITIAL DISTRIBUTION LIST

Combined Arms Research Library
U.S. Army Command and General Staff College
250 Gibbon Ave.
Fort Leavenworth, KS 66027-2314

Defense Technical Information Center/OCA
825 John J. Kingman Rd., Suite 944
Fort Belvoir, VA 22060-6218

Dr. Mark Hull
DMH
USACGSC
100 Stimson Ave.
Fort Leavenworth, KS 66027-2301

LTC Casey Lessard
SOF
USACGSC
100 Stimson Ave.
Fort Leavenworth, KS 66027-2301

Mr. Kenneth Riggins
DCL
USACGSC
100 Stimson Ave.
Fort Leavenworth, KS 66027-2301

Mr. Marc Otte
CDUSM
U.S. Marshals Service
District of Alaska (Tracking Unit)
222 West 7th Avenue, Room 170, Box 28
Anchorage, AK 99513